GERMANY AND ISRAEL

Germany and Israel

MORAL DEBT AND
NATIONAL INTEREST

George Lavy

FRANK CASS
LONDON • PORTLAND, OR

First published in 1996 in Great Britain by
FRANK CASS & CO. LTD.
Newbury House, 900 Eastern Avenue,
London IG2 7HH, England

and in the United States of America by
FRANK CASS
c/o ISBS, 5804 N.E. Hassalo Street, Portland, Oregon 97213-3644

Copyright © 1996 George Lavy

British Library Cataloguing in Publication Data

Lavy, George
    Germany and Israel : moral debt and national interest
    1. World War, 1939–1945 – Reparations 2. Germany – Foreign
    relations – Israel – 1945– 3. Israel – Foreign relations –
    Germany – 1945–
    I. Title
    327.4'3'05694

    ISBN 0-7146-4626-1 (cloth)
    ISBN 0-7146-4191-X (paper)

Library of Congress Cataloging-in-Publication Data

Lavy, George, 1921–
    Germany and Israel : moral debt and national interest / George
Lavy.
        p.   cm.
    Includes bibliographical references and index.
    ISBN 0-7146-4626-1 (cloth). — ISBN 0-7146-4191-X (pbk.)
    1. Germany—Relations—Israel. 2. Israel—Relations—Germany.
3. Germany—Ethnic relations. 4. Holocaust, Jewish (1939–1945)—
Germany—Influence. I. Title.
DD258.85.I75L38 1996
327.4305694'045—dc20   95-12715
                                                              CIP

Typeset by Vitaset, Paddock Wood
Printed in Great Britain by
Bookcraft (Bath) Ltd, Midsomer Norton, Avon

To Phyllis

Contents

Acknowledgements

This book is based on my PhD thesis at the University of London.

I am indebted to the Social Science Research Council for a grant enabling me to undertake a period of study in Germany, to the Royal Institute of International Affairs for allowing me to use their press archives, the Wiener Library in London, the Germania Judaica in Cologne, the Bundestag Library and the archives of the Deutsch-Israelische Gesellschaft in Bonn.

My thanks also to Mr Philip Windsor, Reader in International Relations at the London School of Economics, for his encouragement and help with my research.

G.L.
London, February 1990

Abbreviations

APO	ausserparlamentarische Opposition
CDU	Christlich Demokratische Union
CSU	Christlich Soziale Union
DGB	Deutscher Gewerkschaftsbund
DM	Deutsche Mark
dpa	Deutsche Presseagentur
EC	European Community
EDC	European Defence Community
EU	European Union
FAZ	Frankfurter Allgemeine Zeitung
FDP	Freie Demokratische Partei
GDR	German Democratic Republic
NATO	North Atlantic Treaty Organisation
NPD	Nationaldemokratische Partei Deutschlands
NS	nationalsozialistisch
OPEC	Organisation of Petroleum Exporting Countries
SED	Sozialistische Einheitspartei Deutschlands (the one-time ruling communist party of the GDR)
SPD	Sozialdemokratische Partei Deutschlands
SS	Schutzstaffel (para-military organisation, later part of the regular fighting forces of the Third Reich)
Südd. Ztg.	Süddeutsche Zeitung
UAR	United Arab Republic
UN, UNO	United Nations, United Nations Organisation
US, USA	United States of America
USSR	Union of Soviet Socialist Republics
vol.	volume

Introduction

Sovereign states, in their behaviour towards each other, are normally motivated by self-interest. At least that is the general assumption. The occasions where a government will act for purely moral reasons towards another country are rare. Some say it never happens. It is quite respectable for governments to justify their actions by saying that they were taken in 'the national interest'. And even where there is the rare case of a government acting for a moral reason, it may be reluctant to admit it for fear of criticism by its own people that it is giving something away.

In the case of West Germany there is some controversy about what caused its first governments to act towards Israel in the way it did. But the relationship between the State of Israel, founded in 1948, and the Federal Republic of Germany, established by the victorious western powers in the summer of 1949, is indeed an unusual one. West German–Israeli relations began in the early 1950s against the background of the Hitler regime's extermination of millions of Jews during the Second World War. This act of genocide has had a profound psychological effect on both the Jewish and German peoples which neither has as yet been able to overcome completely. The relations between West Germany and Israel owe their unusual character to the Nazi Holocaust and remain in its shadow to this day.

Shortly after the founding of the Federal Republic the Israeli government lodged a claim against the Germans for material compensation for the horrors committed against the Jews by Hitler's Third Reich. That the West Germans, under their first Chancellor Konrad Adenauer, agreed after protracted negotiations to pay compensation – and indeed a considerable amount – was believed to have been motivated by self-interest: they wanted to ingratiate themselves with the three western powers still occupying their country in order to win concessions from them. But it could be argued that by the time the compensation agreement was being negotiated in the summer of 1952 the West Germans had been accepted by the occupying powers as trusted members of the international community. There would, therefore, be no need for them to demonstrate their atonement for the past by large compensation payment to Israel.

Israel, for its part, considered that the Germans owed to the Jewish people and, by extension to the Israelis, a moral debt of which material compensation was only one factor, to be followed by aid to Israel in the diplomatic, economic and military fields. This was broadly accepted by the West German government but became difficult to implement when in the mid-1950s the Federal Republic, now a sovereign state and no longer occupied, had to conduct its own foreign policy and its government had to respond to the growing tensions in a divided world. Because of the division of Germany the Federal Republic became more deeply involved in the Cold War while Israel was engaged in bitter conflict with its Arab neighbours. Bonn soon became a point of interaction between the two conflicts and had to decide how far it could go in supporting Israel on moral grounds when the national interest dictated other policies.

The object of this book is not to account in detail for everything that has passed between Bonn and Jerusalem over the last 40 years, but to trace the development of West German–Israeli relations in the context of the international pressures exerted on both states during that period, and to show that a case can be made for saying that at important moments in the history of Israel the West Germans supported it for reasons of conscience rather than out of self-interest.

1

First Contacts: Compensation

It was through compensation that the relationship between Israel and West Germany began. The requirement that the Germans must pay compensation to the victims of Nazi oppression, to those, that is, who were physically or economically harmed for reasons of race, creed or nationality, goes back to the period of military government before the creation of the two German states. It was the three western occupying powers that promulgated laws to that effect. But this was a case of compensation directed to individuals and not to the states which had fought the Germans during the war. That came into a different category, usually termed 'reparations'. These were imposed by the Potsdam Agreement, concluded by the four major powers in 1945, well before the State of Israel had come into existence.

Nevertheless a general compensation claim against Germany on behalf of the Jewish victims collectively was made by Chaim Weizmann for the Jewish Agency for Palestine as early as September 1945. Weizmann based his claim on the premise that the Germans had aimed not only at the physical extermination of the Jews but also at the destruction of their religious and cultural heritage and the confiscation of their possessions, individual and communal. The material losses suffered by the Jewish people all over Europe, where thousands of communities were wiped out and their property was confiscated, were estimated at over £2 billion, at that time an astronomically high figure. Although, as Weizmann expressed it, the 'first declaration of war by Germany had been made against the Jewish people'[1] they had not been included among the recipients of reparations at Potsdam or other relevant conferences of the victorious powers. The Weizmann claim established the principle, later endorsed by all parties in the compensation process, that compensation for loss of Jewish institutions or for the property of those who were no longer alive should be placed at the disposal of the representatives of World Jewry and that some of it might be used for the settlement and rehabilitation of Jewish survivors in Palestine.

By the late summer of 1949, when both the State of Israel and the Federal Republic had been established, it would in theory have been possible for a compensation agreement to be negotiated bilaterally between the two governments. But the hatred borne by the Israeli public towards the Germans was such that no government of Israel could have made contact with any German authority. This hardly requires explanation when it is considered that at the time a large, and increasing, proportion of the Israeli population was made up of Holocaust survivors and those who had lost relatives at the hands of the Nazis during the war. Israel at that time did not recognise the existence of either German state and did not wish to communicate with any Germans.

That is why first offers by the Federal Government of material compensation for the mass killings and for the confiscation of Jewish property were either ignored by Israel or rejected as 'blood money which after what had occurred no self-respecting nation could accept'.[2] As late as September 1950 a prominent Israeli politician protested at the meeting of the Interparliamentary Union in Istanbul that it was an 'insult to any honest and decent human being to have to negotiate with Germans as if the Germans were free of all that had been done in their name and especially to the Jewish people'.[3] In the Federal Republic too, it was by no means clear that the majority were anxious to re-establish relations with the Jews. The West German public, emerging from years of starvation and economic ruin, had other priorities. Many felt they had had no part in the destruction of European Jewry and that no blame therefore attached to them. If the government and a minority of public figures expressed the view that the German people owed a heavy moral debt to the Jews, that belief was not shared by the majority.

The question of a future relationship between Israel and West Germany began to arise only when the government of Israel drafted claims against the Germans and Konrad Adenauer, the first Chancellor of the newly founded West German state, accepted in principle that such claims were justified and negotiable. The governments of the two countries, for different reasons, in the end saw the need to bring about a rapprochement. International political changes were important, if not exclusive, in this development. The International Political System which became established in the late 1940s and early 1950s affected virtually all countries of the world, forcing many to take sides in the power struggle developing between the two superpowers. As will be seen, both the Federal Republic and Israel were ultimately drawn into the East–West conflict. It was this conflict that indirectly helped the two governments to move towards each other, even though public opinion in Israel resisted this development for some time.

At this stage it is expedient to recall that the State of Israel was created by a majority vote of the members of the United Nations and that the

2

Soviet Union was of this majority. Three days after the founding of the state the USSR accorded it recognition. For some time thereafter all Soviet Bloc states gave economic and even military support to the new state which was then fighting off an invasion by its Arab neighbours. In Israel itself there was considerable sympathy towards the Russians, which may not have been unconnected with the large Russian element in Israel's population and the political and cultural influence former Russian immigrants had in the country.[4] In this situation it was not difficult for the young state to assert its neutrality. On 31 January 1949 David Ben-Gurion, the Prime Minister, stated that Israel would co-operate with both the USA and the USSR to promote world peace. Commercial agreements were signed between Israel, Hungary, Poland and Czechoslovakia.[5] Israeli neutrality seemed to be holding despite the increasing whirlwind of the Cold War.

Israel therefore had no difficulty in establishing itself in the international community. It was accepted by both the West and the Communist Bloc and became a member of the United Nations in May 1949. Only the Arab countries remained implacably hostile to the very existence of the new state. They did not accept the right claimed by the Jewish people of recovering their ancient homeland and disagreed that large-scale Jewish immigration into what had long been a mainly Arab populated Palestine was justified by the need to find a haven of refuge for the victims of Hitler's extermination policies. This hostility received some sympathy in the newly emerging states of the Third World but, at first, no support from the great powers that dominated the international scene in the late 1940s. Even the fact that Israel had gained territory at the armistice following the war of 1948–49 with its Arab neighbours over and above that which had been envisaged by the United Nations partition plan for Palestine did not create any international problems.

By contrast, the newly founded Federal Republic of Germany at first had serious difficulties of legitimation. It was created by the West against the strong opposition of the Soviet Union, which had been one of the signatories of the Potsdam Agreement concluded after the Second World War. This agreement had been intended to establish the future status of Germany as an undivided country, though occupied and governed for a time by the victors. The creation in 1949 of the West German state, by which the western powers had sought to serve their security needs at a time of deepening conflict with and fear of the Soviet Union, was technically in breach of the Potsdam Agreement. It was therefore not recognised by the communist countries. Even in the rest of the world and in West Germany itself the Federal Republic was not at first regarded as a fully fledged state; its founding was considered as a provisional solution, to be maintained only until the differences between East and West had been overcome. The three western occupying powers continued to exercise sovereign rights over the new state: they conducted all its external relations and were responsible for its defence. Above all, the occupation was maintained with significant rights of intervention in the internal affairs of the Federal Republic remaining with the occupiers. The Germans were not yet trusted and had to prove to the world that they had eradicated all traces of Nazism and militarism and purged from their minds all ideas of world domination, that they were willing and able to develop a stable and democratic society and live in peace with their neighbours. The picture of the Federal Republic during the first few years of its existence was therefore one of a state existing on sufferance, without recognition or formal relations with any states except the occupiers, and unable to be a member of the United Nations or any other international organisation.

In practice, however, the Federal Government did have a few cards to play, which after a surprisingly short time enabled it to improve the quality of life of the people and the international status of the country. The main card was the Federal Republic's potential usefulness as an ally of the West in the deepening Cold War between the superpowers: the new republic, once it was seen to be fulfilling its obligations of establishing a stable democracy and co-operating with the West, could become a useful participant in the strengthening of western Europe in the face of real or perceived threats from the Soviet Union. Another card was the phenomenal economic recovery of the Federal Republic in the mid-1950s, which greatly increased its importance and prestige, all the more so since several of the western European states were experiencing serious economic difficulties at that time. West Germany's economic strength also made the country a desirable trading partner and financial benefactor to many Third World countries, about which more will be said in later chapters. But the most important asset, the country's potential usefulness to the West, was quickly noted by Adenauer, whose approach from the moment he came to power was to trade the co-operation of his country with the western occupiers against concessions by them which would lighten the burden of occupation and ultimately remove it altogether, thus restoring the country's sovereignty. Adenauer had little difficulty in satisfying the occupiers, with whose outlook and political aims he broadly agreed. He had suffered under the Hitler regime for his opposition to Nazism, and as a strong believer in the moral regeneration of the German people he soon gained the confidence of the western governments. A strong anti-communist, he had since the end of the war foreseen the division of Europe and expressed the belief that in such a situation the areas of Germany not occupied by the USSR would become 'an integral part of western Europe'.[6]

For Israel, too, the situation soon began to change, though mainly to its disadvantage. On 31 May 1950 Prime Minister Ben-Gurion reaffirmed once more his country's desire to remain neutral. Two related factors, one economic and one political, soon made the holding of this position impossible. The first was the rapid deterioration of the Israeli economy due to the mass immigration of refugees from Europe, mostly survivors of the Nazi Holocaust, to which soon had to be added many Jews fleeing from the wrath of Israel's Arab neighbours. Of the funds needed to settle 600,000 refugees in the country, only about a third could be found in Israel itself;[7] the remainder had to come from elsewhere, mainly from the United States. Israel was thus forced to rely increasingly on the West, especially on the Americans, but this economic dependence exacted a political price – some changes in Israel's unique socialist economy. By the autumn of 1950 the cabinet announced economic reforms relaxing governmental controls on business and making other concessions to private enterprise. The second factor was that the Russians were switching their sympathy and support from Israel to the Arabs, especially in the United Nations. At this time also Stalin began a wave of persecution of the Jews in the USSR and Jewish emigration from there to Israel was stopped. Faced with this situation as well as with the hostility of its Arab neighbours Israel had to avoid complete isolation. When in the summer of 1950 the outbreak of the Korean War added a new military dimension to the East–West conflict and heightened tension throughout the world, a vote by the Israeli parliament approved the Security Council's resolution in favour of South Korea. It was the first important signal of Israel's departure from neutrality.

Israel's internal problems and increasing economic and political identification with the West were soon to affect its relations with West Germany. In the first instance it became increasingly difficult for the Israeli government to reject the financial compensation offered by the Federal Government for the atrocities of the Third Reich. On the political side the Federal Republic was becoming the preferred ally of the USA in western Europe and was forging closer links with France, of which it had become a partner in the newly established European Community for Coal and Steel by 1952. If Israel was to avoid the danger of isolation by establishing close relations with western Europe, both France and the Federal Republic would occupy key positions in this relationship. Despite Israel's reluctance at first to entertain any relations with the West Germans, these were factors which its government could not ignore for long.

As was hinted earlier, the West German government accepted the need to tackle the problem of its moral standing vis-à-vis the Jews. Two months after taking office in the summer of 1949, Chancellor Adenauer, in a first interview with a Jewish journalist, indicated that the Federal Republic was prepared to pay compensation for material losses inflicted on the Jews by the Hitler regime. This interview also contained a first hint that the State of Israel would become involved. Adenauer at this stage offered goods for reconstruction to the value of DM10 million to Israel 'as a sign that the injustice inflicted on Jews throughout the world by Germans must be made good'.[8] At that the Jewish leaders, mindful of the far larger claim made in 1945 by the Jewish Agency, were unimpressed, while the Israelis ignored the offer completely. But so soon after the end of the war the Federal Republic's economic position did not permit anything more than a token offer and this first gesture of Adenauer's therefore remained without a sequel. During 1950 and 1951 the Israeli government, at that time still unwilling to have any dealings with any German authority, addressed several notes relating to compensation to the western occupying powers. Only one of these, sent in March 1951, is of interest here. It moves away from the Jewish Agency's estimate of the value of destroyed Jewish property in Europe and instead bases its claim on the cost of receiving the remnants of European Jewry who had immigrated to Israel up to that time. The estimate of Jewish property was in fact difficult to substantiate while the burden of immigration could be quantified with some accuracy. It was stated to be $3,000 for each of 500,000 European immigrants to date, making a total claim of $1.5 billion. Two-thirds of this was to be paid by the Federal Republic and one-third by the German Democratic Republic. Such a claim was not only felt to be justified but likely to make sense to politicians and world opinion generally.

There was considerable disagreement within Federal Germany about the need to offer compensation of the kind envisaged by the Israeli note of March 1951. A few unrepentant supporters of the Hitler regime considered that nothing was owed to the Jews by any German. A majority of those who in the post-war years were prominent in public life, however, had accepted the obligation to compensate those Jewish victims who had suffered persecution and were still alive. The leaders of all political parties had expressed themselves unequivocally to this effect, both inside and outside the Federal Parliament. Compensation to the State of Israel was another matter. There was no claim in international law for 'reparations': the State of Israel had not come into existence until after the war. Many West Germans, therefore, used the legal argument of the non-existence of Israel at the time of the Nazi crimes to oppose the Israeli claim.[9]

It also troubled the Western Allies, to whom the note had been addressed and who at that time were trying to build an economically

6

strong West Germany as part of their plan to strengthen western Europe against what they perceived as a Soviet threat. The Americans were bearing the major burden of reconstruction in western Europe and other parts of the world and there was concern in Washington that economic recovery in the Federal Republic, which was about to undertake heavy new international commitments apart from possible compensation to Israel, would be impeded. Although by 1951 West German economic recovery was proceeding well, no one inside or outside the Federal Republic could have foreseen the phenomenal strides which that country was to make in due course. For reasons which will become clear before the end of this chapter the US government refused to deal with the Israelis' demands and told them that any agreement of the kind suggested would have to be negotiated directly with the Federal Government.

In the course of 1951, following the Israeli note, a number of developments of great importance for West Germany took place. At the end of May the Federal Republic, together with five other western European states, signed the treaty for the European Defence Community, and with France, the United Kingdom and the USA the Contractual Agreements, known in Germany as *Deutschlandvertrag*. Under these treaties the occupation of the Federal Republic would end and the country could become a sovereign state, but at the same time it would be committed, on joining the EDC, to establish armed forces and a tactical air force, providing 500,000 men as a contribution to a European army. The German financial burden would be in the nature of DM11.5 billion.[10] Later that year the Conference of Jewish Material Claims against Germany – usually known as the Claims Conference – was set up, embracing a large number of Jewish organisations throughout the world and with the objective of co-ordinating Jewish claims against Germany, including the Israeli claim. Finally there was the London Debt Conference to negotiate the amount of the pre-war German debt which the Federal Republic, as the successor of the former German Reich, had expressed a willingness to settle. Despite all these heavy commitments the Adenauer government made a declaration in the Bundestag in September that it was 'ready, jointly with the representatives of Jewry and the State of Israel, which has received so many homeless Jewish refugees, to bring about a solution of the problem of material restitution ...'.[11] This was the first public utterance by the West German government that it was prepared to go further than the token offer made to Israel in 1949.

The declaration was cautious. It stressed the civil rights aspects of the constitution, which gave equality to all citizens irrespective of race or creed. It pointed to the need to educate people towards the spirit of human and religious tolerance and gave warning that anti-Semitic or racist propaganda of any kind would be severely punished. It emphasised the steps that had already been taken to compensate individual Jews for material damage done to them by the Nazi regime; and only at the end made its statement of intent, while warning at the same time that the amount of compensation had to be considered in the light of the strength of the economy, having regard especially to the need of the government to provide for the many German war victims and refugees from Germany's lost eastern provinces.[12] Defensive though it was in its tone, this declaration formed the basis for serious negotiations. It was followed in December by a letter from Adenauer to Dr Nahum Goldmann, Chairman of the Claims Conference, stating that in the view of the Federal Government the time had come for negotiations to start. He added that his government agreed to accept the claims of Israel, as stated in the Israeli note of March 1951, as a basis for discussion. The declaration by the Federal Government was approved by a large majority in the Bundestag.

In West Germany this new move met with a rather mixed reception. In Israel the reaction was predictably violent. The Israeli government had refrained from negotiating officially with the Germans, though unofficial contacts, which were kept secret, had been made. The discussion whether Israel should be a party to a compensation agreement with West Germany had been kept discreetly private in official circles for fear of a public outcry. In the press and among the public, on the other hand, the subject of compensation by the Federal Republic had always raised strong emotions, and the idea of accepting them had on the whole been rejected. Despite this the Israeli government decided to respond to the German offer, but only after first obtaining the agreement of the Knesset for negotiations to start. The result was pandemonium. There were violent scenes outside the Jerusalem parliament building, where protest demonstrations were held and stones were thrown, while inside repeated interruptions caused the temporary abandonment of the session before a vote in favour of negotiations was finally obtained.

The Israeli government at this time had urgent reasons for wanting to establish contact even if the public were not yet ready for it. Israel was not a wealthy country, has few natural resources and was engaged in establishing a modern agriculture where land was largely desert. The settling of refugees was in full swing. The economy was in urgent need of an injection of finance from outside and the German offer was therefore too tempting to be rejected out of hand. The government was also mindful of Israel's international position, that of a country surrounded by hostile neighbours and with few friends elsewhere in the world. The Federal Republic, by now a founder member of the European Community for Coal and Steel and about to join the European Defence Community, was likely to play one day an important role in a resurgent western Europe. If these considerations were not readily understood by the general public, there could be no question of the government simply giving in to the irrational reactions of the people, however great may have been the sympathy of members of parliament and of the cabinet with the feelings of the many who had suffered from Hitler's 'Final Solution' and who were not yet ready to dissociate the new democratic Germany from its Nazi past. In the Knesset debate, therefore, the Prime Minister skilfully countered the arguments of the main opposition party, the Herut, that it would be iniquitous to strike any deal with the Germans, by insisting that Israel in fact had a duty to demand compensation for 'the wealth that had been plundered by the Nazi regime'. After two days of debate, late on 8 January 1952, despite the demonstrations outside, the Knesset gave the green light for the compensation negotiations to start.[13]

The events which began on 21 March 1952 at Wassenaar, a small town near The Hague, and ended in the signing of the 'Restitution Agreement' at Luxembourg on 10 September have been described in detail elsewhere. Some of the problems that arose should, however, be mentioned here. The principle of compensation, based on the cost to Israel of receiving the survivors, had been established before the start of the negotiations, but the amount became the subject of tough bargaining, with the Germans fiercely resisting the claim of $1.5 billion, as formulated in the Israeli note of March 1951. The German resistance was explained by the economic situation in the Federal Republic, which was still in the process of rebuilding seven years after the end of the war, as well as by the commitments mentioned above. But the matter was complicated most of all by the London Debt Conference, in session at the same time and endeavouring to fix the German pre-war debts to the Allies.

Whereas the negotiations concerning the European Defence Community and the contribution the West Germans would have to make to it were almost completed, the London Debt Conference was going on when the Wassenaar talks about compensation started, and the German debt had yet to be fixed. The importance to the Germans of coming to terms with their pre-war creditors was that it would re-establish West German credit-worthiness in the eyes of the West, which was seen as a prerequisite to a full return of the Federal Republic to the world markets and thus to complete economic recovery. It was considered by many Germans to be more important than the restoration of the Germans' 'moral credibility', which would result from a compensation agreement. As the Federal Government was anxious to make a generous gesture to the erstwhile victors, it seemed that the size of the compensation 'cake' would depend on what was left after the Allied debts had been paid.

9

Adenauer's cabinet was not only divided on the issue of how much compensation the country could afford to offer to Israel, but the Federal Finance Minister, Fritz Schäffer, declared categorically, and repeated this throughout the negotiations, that there were no funds available for compensation.[14] The German side at first, therefore, made an offer to Israel which fell substantially short of the amount demanded in the Israeli note and accepted by the Germans as a basis for negotiations. This offer was rejected by the Israelis, whereupon the Germans decided to delay their final decision until the outcome of the London Debt Conference could be more clearly foreseen. Several attempts by the Germans to obtain American financial aid, which might have overcome the difficulties, were turned down by the United States government.[15] In April the Wassenaar conference reached deadlock and the talks were suspended.

There is still some controversy as to the part played by the Western allies in the way the Wassenaar Conference came about and how it was restarted after the breakdown. No direct pressure seems to have been put on the Federal Government before or during the negotiations by the United States or any other western government. The Americans, while sympathetic to the idea of German compensation and eager that the Federal Government should come to terms with Israel, were concerned that one or several consequences might arise out of large German compensation payments: the Germans might be forced to scale down the debt repayments; they might be unable to meet their rearmament commitments within the EDC; they might have to ask for more financial aid. The United States were not willing to pay – as it were – part of the German compensation bill.[16] Neither were they inclined to impose excessive financial burdens on the Federal Republic at a time when they wanted it to become a partner in the defence of western Europe against possible Soviet aggression. In addition, the US government was unwilling to antagonise the Arabs by being seen to become too involved in helping the Israelis obtain large sums in compensation at a time when it was trying to organise a Middle East alliance against the Soviet Union, which, it was hoped, would include Arab states.[17] More will be said about this in the next chapter. There were thus enough reasons for the Americans to be reluctant to press the Germans, especially over the amount of compensation. This explains why the US government in its reply to the Israeli note of March 1951 refused to act as broker over compensation, their refusal to provide loans to Bonn, as suggested by the West German Finance Minister and Adenauer's financial advisers, and their reluctance to comply with several Israeli requests to apply pressure on the German negotiators to meet their demands.

With the US government apparently content to watch the negotiations with 'sympathy and interest', the final decision was left to the West Germans. Adenauer was under severe pressure, firstly because of his work load, since he had to attend to the other negotiations in which his country was engaged, but also because of the divisions in his cabinet, especially the stubborn resistance led by his Finance Minister. He could not ignore the arguments of the opposition that compensation at the rate the Israelis were demanding was beyond the means of the Federal Republic and could lead, if pursued, to serious economic difficulties. But there were supporters of the Israeli case. In the cabinet these were Ludwig Erhard, the Economics Minister; Walter Hallstein and Herbert Blankenhorn, Secretaries of State respectively in the Chancellor's Office and the Foreign Ministry; and, outside, the two German negotiators at Wassenaar, Franz Böhm and Otto Küster. The two negotiators resigned in protest over the government's stalling.[18] Behind the scenes, as an unofficial observer, was Nahum Goldmann, who urged the Chancellor to accept the Israeli demands.

Thus it was in the end Adenauer who, on the advice of Goldmann,[19] decided to give in to the Israelis and to do so without waiting for the London Debt Conference. The Germans finally agreed to pay DM3 billion to the State of Israel and DM450 million to the Jewish organisations represented at the Claims Conference. The payment would be mainly in goods and effected over a period of 14 years. The 'Restitution Agreement' was signed in Luxembourg on 10 September 1952. Both inside and outside the cabinet Adenauer stressed the moral need for the Germans to offer adequate compensation for the atrocities committed by the Third Reich against the Jews, but he also emphasised the harm a failure to reach agreement could have caused to the reputation of the German people, which in turn might have impeded the progress of the Federal Republic towards full integration in the western world. Adenauer may have over-estimated the danger of western hostility in case of failure, perhaps because American public opinion had, at the time of the suspension of the Wassenaar negotiations, become very critical of the Germans, as was shown by hostile comments in the American press.[20] But, more likely, he may have deliberately overstated the political dangers of failure in order to convince the German public and those of his opponents in the cabinet, who may have rated *Realpolitik* higher than a moral imperative.

In asking the question why the Federal Republic committed itself against heavy odds to making the very large compensation payments to Israel, it would be tempting, in a study of an international relations topic, to invoke only the national interest and to relate this aspect of Adenauer's politics exclusively to a desire to impress the western occupiers of his country in order to gain political advantage, as he did on other occasions.

11

It is true, of course, that the Germans were under close scrutiny by the governments and peoples of their former enemies, as was demonstrated by public reaction in the United States when the Wassenaar talks looked as though they might founder. A failure of being seen to clear their reputation could have had adverse consequences, especially in western Europe, where the American proposals to let the Federal Republic rearm and participate in the defence of Europe were met at first with misgivings. But such an arrangement was in the early 1950s hardly any longer in doubt, since the government of the United States, which had taken over the leadership in the West, insisted on the Germans being given such a role. The only condition, that the West Germans proved themselves as reliable allies by abjuring Nazism, establishing a sound, stable democracy and co-operating with the West, had already been largely fulfilled. Bonn did not therefore need to impose on the country the heavy burden of compensation payments to Israel in order to achieve the trust of the western governments. For although Adenauer said later that leading world bankers, especially in the US, would have exerted a negative influence on the success of the London Debt Conference if the negotiations with Israel had failed,[21] the American government was, for the reasons stated above, only lukewarm in its support for the Wassenaar negotiations.

The largest share of the credit for achieving the Luxembourg Restitution Agreement must therefore go to the Chancellor, Adenauer, and some of his closest aides. They acted sincerely out of their consciences and in that made themselves into the conscience of the people. For Adenauer political advantage only played a subsidiary part. There are other examples during Adenauer's lifetime of his support for Israel, of decisions which could not have been described as having been dictated purely by the national interest. Adenauer may have been conscious of a need to impose on the German people a burden which would sharpen the awareness of the iniquities of the Nazi regime as well as help the people by its sacrifice – and such it appeared at the time – to alleviate the national conscience and restore its self-respect. The Treaty was ratified by both houses of the West German parliament in March 1953.

After the successful conclusion of the Luxembourg Agreement it was even more its rapid ratification and conscientious implementation by the West Germans – against the background of mounting economic and political pressure by the Arabs on their government – that established a lasting relationship between the Federal Republic and Israel. Both states have referred to it as a special relationship, determined by recent history: the horrors committed by the Third Reich against the Jews. West German politicians and other public figures have frequently spoken of a moral debt owed by the Germans to the Jews and in particular to Israel, while the Israelis have said that the West Germans have accepted a moral responsibility towards them and their state. But not surprisingly, this relationship soon became affected by the strong pressures and even occasional storms engendered by the International Political System.

NOTES

1. Chaim Weizmann: *Note to USA, UK, USSR and France*, 20 Sept. 1945.
2. Rolf Vogel: *Deutschlands Weg nach Israel*, Seewald Verlag, 1967, p. 19.
3. A.L. Easterman: *Monument without Epitaph*, World Jewry, 1962.
4. See Amos Elon: *The Israelis. Founders and Sons*, Harmondsworth: Penguin, 1983, chs. 1–7.
5. *Middle East Journal*, Spring 1949 and Spring 1950.
6. Letter dated 31 Oct. 1945 by Adenauer to the mayor of Duisburg. See K. Adenauer: *Erinnerungen*, vol. 1, Frankfurt a.M., p. 40.
7. *Middle East Journal*, Winter 1951.
8. Adenauer interview with Karl Marx, ed. of *Allgemeine Wochenzeitung der Juden in Deutschland*, 11 Nov. 1949.
9. Felix E. Shinnar: *Bericht eines Beauftragten*, Tübingen: Wunderlich Verlag, 1967, p. 24.
10. F. Roy Willis: *France, Germany and the New Europe*, New York: Oxford University Press, 1968, p. 136.
11. Shinnar, op. cit., p. 28.
12. Shinnar, op. cit., p. 28.
13. Shinnar, op cit., p. 31. The vote was 61 in favour, 50 against, with 4 absences and 5 abstentions.
14. Michael Wolffsohn: 'Globalentschädigung für Israel und die Juden? Adenauer und die Opposition in der Bundesregierung', in: L. Herbst and C. Goschler (eds): *Wiedergutmachung in der Bundesrepublik Deutschland*, Oldenbourg, 1989. Kai von Jena: 'Versöhnung mit Israel? Die deutsch-israelischen Verhandlungen bis zum Wiedergutmachungsabkommen von 1952', in: *Vierteljahrsheft für Zeitgeschichte*, vol. 34, no. 4 (Oct. 1986).
15. Wolffsohn: op. cit. Wolffsohn: 'Das deutsch-israelische Wiedergutmachungsabkommen von 1952 im internationalen Zusammenhang', in: *Vierteljahrsheft für Zeitgeschichte*, vol. 36, no. 4 (Oct. 1988).
16. N. Balabkins: *West German Reparations and Israel*, New Brunswick: Rutgers University, 1971, p. 126.
17. Wolffsohn: 'Wiedergutmachungsabkommen von 1952', op. cit.
18. von Jena: op. cit. Prof. Böhm later withdrew his resignation, unlike his colleague, Otto Küster.
19. von Jena: ibid.
20. Norbert Frei: 'Die deutsche Wiedergutmachungspolitik gegenüber Israel im Urteil der öffentlichen Meinung der Meinung der USA', in: *Wiedergutmachung in der Bundesrepublik Deutschland* (Herbst and Goschler, eds), Oldenbourg, 1989.
21. R. Vogel, op. cit., p. 42.

2

The International Political Scene in the Fifties

It is impossible to understand West German–Israeli relations without stating how the politics of these two countries interact with the strains and stresses elsewhere in the world. Although from what has been said so far this relationship may appear superficially to be a purely bilateral one, based only on moral considerations, both countries were involved in conflicts elsewhere, the State of Israel since the day it was proclaimed, the Federal Republic increasingly as it regained its sovereignty, that is, its right freely to conduct its own foreign policy. How external pressures were brought to bear on the negotiations and the ratification process of the Restitution Agreement has already been hinted at. But the commencement after the Paris Agreements in 1955 by the Federal Republic to play an international role again, however modest and passive at first, coinciding with changes of superpower activity in the Middle East, put serious strains on the new West German–Israeli relationship. The Middle East must be regarded today as the most effervescent and potentially dangerous regional sub-system of the now global International Political System. It extends over the heartlands of Islam, that is, all Arab areas, including those of North Africa, Turkey, Iran, Pakistan, Afghanistan and south into the Horn of Africa. Its conflicts are mainly of local origins promoted by local rivalries, power politics and ideological differences, complicated in the late 1940s by the appearance on the scene of a newcomer, the State of Israel. As was the case with older multi-polar systems, this one is characterised by frequent changes of alignments, the formation and reformation of alliances, but also by attempts to create some kind of balance of power.[1] In 1955, however, both superpowers became involved but they used the Middle East mainly as a means of achieving their global aspirations and of playing out their own rivalries. This has caused them to intervene more and more in the struggles of the Middle East, but without being able to dominate it to any great extent. The difficulties of the Federal Republic in its relations with the Middle East sprang from three main causes: that because of the burden of Germany's past

14

it became closely tied to Israel with which it established a special relationship; that it also actively promoted good relations with the Arab states, locked in bitter conflict with Israel; and that it became vulnerable as a result of the division of Germany and its aspirations of reunification, which brought it into conflict with the Soviet Union and thus into the Cold War.

THE MIDDLE EAST

The history of the Middle East as an active regional sub-system goes back to the end of the First World War when the division of the Ottoman Empire led to the creation of a number of new nominally independent states whose boundaries were drawn not so much from ethnic, cultural or historical considerations but rather as a result of rivalries between local dignitaries or between the victorious western powers themselves. Since that period, four major conflicts, all of them separate in origin but increasingly interacting, can be identified. It started with the Arabs' resistance to the western European colonial powers which had created their states out of the ruins of the Turkish Empire and had occupied or dominated them. To this was added, after the Second World War, the Arab–Israeli conflict which has caused international tension since the day the State of Israel was founded in 1948. Then, by 1955, the USSR, after first maintaining a low profile in the area following the end of the war, began to take an active interest by giving diplomatic and military aid to some Arab states in their struggle against western domination. Because of the deep political and economic involvement of the western states, which now included the USA, in the Middle East, Soviet entry into Middle Eastern politics meant that the Cold War extended into this region. Finally there are the many intra-Arab rivalries leading to varying alignments and conflicts which, though rarely erupting into open warfare, had an effect on the eastern Mediterranean area, including Israel and its relations with other countries. The first three of these conflicts played some part in the development of relations between the Federal Republic and Israel, and some attention must therefore be given to them. The intra-Arab conflicts, often due to internal upheavals in some of the countries, are of relevance mainly in that they have added to the general volatility of the region.

The occupation of much of the Middle East by Great Britain and France had at first raised hopes among Arab leaders that their peoples would now obtain independence and that a new Arab revival after centuries of dependence and eclipse was on its way. Although western liberalism does not easily accord with the Islamic philosophy of life it had many admirers in the Muslim world. The disillusionment became all the greater when it was found that liberal principles were not being applied to the newly 'liberated' Arab peoples and the Arabs soon began to feel that they had simply exchanged one form of foreign domination for another.

Great Britain's role in the Middle East was not that of the traditional colonial power. But although the Arab countries it had freed from Turkish rule had become nominally independent they were tied to the United Kingdom by treaties of a military and economic kind which greatly reduced their freedom of action. As long as Great Britain needed to safeguard the sea and land routes to India, the Far East and Australia, it had an interest in exercising control over certain Middle Eastern countries to make sure that no rival gained a foothold there. But British control could be maintained only as long as the feudal princes and the small ruling class on which their power rested were willing to extend to their imperial masters a measure of co-operation. This they were on the whole prepared to do because of a community of economic and commercial interests with the British. But they were increasingly challenged by a new generation of Arab nationalists who wanted to free their peoples both from the domination of foreign powers and from the ruling classes who had become their allies. To protect themselves against this threat the interested western powers found it increasingly necessary to give active, sometimes military help to the ruling classes and their governments at a time when these no longer enjoyed popular support. In time, such interference in the internal affairs of their Arab client states created bitterness and in some cases contributed to the overthrow of the old regimes by popular revolutions.

At the end of the Second World War Great Britain was for a short time the strongest power in the Middle East, virtually unchallenged by any rivals. In the late 1940s, however, partly because of increasing nationalist ferment in some Arab countries, partly because of economic difficulties at home, Britain's power in the eastern Mediterranean declined. The USA now increasingly took over the position the UK had held in the area. British imperial lines of communication were no longer the main reason for western hegemony: the concern of the US government was to prevent the Soviet Union from establishing itself there, especially since the western industrial nations were becoming more dependent on Middle Eastern oil.

With the increasing American influence tactics also changed. The United States were at first reluctant to become too deeply involved in Middle Eastern affairs, leaving much of the onus still on the British. They did however encourage the formation of defensive alliances – without themselves participating in them – among the Middle Eastern states. The inducements for joining these alliances were economic and military aid.

But this new policy was successful only in the mainly non-Arab 'Northern Tier' of Middle Eastern states comprising Turkey, Iran, Afghanistan and Pakistan. These countries, together with Iraq, the only Arab state, combined together in the Baghdad Pact, of which Great Britain, though not the United States, also became a member. But the other Arab states refused to join. One important reason for this was the growing anti-West feeling in the Arab world. In the early 1950s revolutions occurred in several Arab countries during which the old-style monarchies were replaced by republican governments whose aim was to liberate the Arab peoples from foreign domination. These new regimes rejected the American alliance policy as a new form of western domination.

These changes naturally created complications for the West, not only because of the removal of some of the pro-western monarchies, but also because there was now a new division and indeed tension among the Arab states. The revolutionary regimes became officially non-aligned but because of their implacable hostility to western imperialism often showed sympathetic leanings towards the Soviet Bloc. The most important of these was Egypt. There the revolution of the 'Free Officers' had ousted King Farouk in 1952 and paved the way for Abdel Gamal Nasser to eventually take over the leadership. The advent of this new regime, with its claim to Arab leadership in the cause of Arab unity and the struggle for liberation from western domination, not only challenged western policies but led to increased Egyptian hostility towards Israel.

Since 1948, the year when the Jewish State came into existence, the conflict between the Arab peoples and the West has been compounded by the conflict between Israel and its Arab neighbours, the second of the four conflicts afflicting the Middle East. As soon as the British mandate over Palestine was terminated and the new state was proclaimed, all the Arab states on or near its borders invaded it but their attempt to destroy it was defeated. After that time no formal peace was concluded between Israel and the invaders and the relations between them could at best be described as an uneasy armistice, interrupted by four major military campaigns and innumerable smaller border incidents. The Arab attitude has been that the Israelis are foreign intruders who have taken away land that had belonged to Arabs and driven many of its inhabitants into exile. Jewish refugees, so the arguments ran, were being settled in Palestine, thus creating a new, this time Arab, refugee problem: the Arabs rather than the Jews were ultimately made to pay for Hitler's crimes, which had not been directed at them and with which they had nothing to do.[2] In the sense that Great Britain had since the Balfour Declaration in 1917 established a national home for the Jews in Palestine and the United States were supporting the State of Israel, the Arabs' struggle against Israel became in

Arab eyes closely linked to their fight for freedom from western domination. No Arab state could come to terms with Israel and a state of belligerency continued to exist.

Because of this situation the State of Israel had to bear an unnaturally large defence burden. The problem was not only a financial one, though this was serious enough: there was an acute difficulty in obtaining the right sort of weapons that the Israeli armed forces needed. Both the United States and some western European states were reluctant to supply arms to Israel for fear of offending the Arabs. The Americans' failure to bring most of the Arab states into their alliance system made it imperative that they should restrict their supply of arms to Israel lest they should appear to favour the latter militarily. Though successive United States administrations have declared the survival of Israel a primary American interest they had to take account of the Middle Eastern situation as a whole. That is why Israeli interest in arms purchases was increasingly directed to other areas, notably western Europe. This will be more closely examined in the following chapters. Meanwhile, by 1954, there were signs that after years of abstention the USSR was beginning to take a renewed interest in Arab affairs.

Russian interest in the eastern Mediterranean had already been intense in the nineteenth and early twentieth centuries, long before the Revolution of 1917. After 1945, therefore, there was concern in the West that the expansion of Soviet influence might be extended to the Middle East. Communist threats to Greece and Turkey and the Soviets' reluctance to withdraw their troops from northern Iran after the war seemed to indicate a coming Soviet thrust southward. But, whether as a result of British and American counter-measures or preoccupation with Europe or displeasure over the reactionary nature of the then mainly monarchical Arab regimes, Soviet pressure relaxed and there was no evidence of a serious Soviet involvement in the Arab world. As time went on the emergence of nationalistic, strongly anti-western governments gave the Soviets a chance to gain favour among some Arab states. The Middle East conflicts thus began to sharpen. Soviet diplomatic activity now increased and was vigorously pursued from 1954 when their support for the Arab cause against Israel in the United Nations was intensified.

An opportunity was afforded by the conclusion of the Baghdad Pact early in 1955. The Pact had caused an uproar in many Arab countries where it was seen, because of the membership of Iraq, as an attempt to divide the Arab world and consolidate weakening British imperialism. The adherence of Iraq, one of the remaining bastions of pro-British policies, was particularly resented by President Nasser of Egypt, who actively sponsored a counter-alliance with Syria, Saudi Arabia and Jordan.

But anger was not confined to governments; popular support for Arab independence was growing all the time and was vigorously fanned by propaganda, especially from Egypt. The power that benefited from this row was the Soviet Union.

The occasion for the first direct initiative by the Soviets was provided by the Bandung Conference of Non-aligned Nations, held one month after the British signing of the Baghdad Pact. On the eve of the Conference and timed no doubt to make an impact on its participants, a Soviet government declaration, clearly aimed at the Baghdad Pact, criticised 'alliances that endanger peace, security and the national interest of states' in the Middle East. The declaration correctly anticipated the tenor of the Bandung Conference which was anti-colonialist, anti-western as well as anti-Israel. It made a strong impression on many of the delegates and in the view of some Middle Eastern statesmen now established the USSR as an active champion of – ironically – non-alignment and of the Arab nationalist cause. In particular it is thought to have encouraged President Nasser to seek Soviet aid in an attempt to lessen his country's dependence on the West. Egypt is the most populous and potentially the most power-ful Arab country. Nasser, whose flamboyant style and brinkmanship politics were to dominate the Middle East for nearly two decades, had achieved during his short time in power a notable diplomatic success against a western power: the conditional withdrawal of British forces from the Suez Canal area which they had occupied under an Anglo-Egyptian treaty since 1936. Though his aspiration to Egyptian leadership in the Arab world had caused misgivings among the conservative Arab rulers, his diplomatic achievement vis-à-vis the British was regarded as a victory for Arab independence and therefore much admired. The encouraging noises made by the Soviet government held out new prospects for further exploits against the West. The bomb-shell came in September 1955 when Nasser announced that he had succeeded in obtaining arms from the Czechoslovak People's Republic, the first time that a Soviet Bloc country had agreed to supply weapons to an Arab state. He had wanted these for use against Israel. The West, concerned about the arms balance and angry with Nasser over his opposition to the Baghdad Pact, had refused to supply them. But now the delivery of arms to an Arab Middle Eastern country by a member of the Soviet Bloc clearly indicated that the USSR had become actively involved. The Cold War – as the third conflict in the Middle East – had come to this area.

Some Arab statesmen now welcomed a Soviet role and were tempted to turn to the Soviet Union for help against the West. The attraction in Arab eyes was not Soviet Communism but the fact that the USSR had not been directly involved in Middle Eastern affairs before; it was free from the charge of colonialism. By contrast, the western powers seemed to be omnipresent and interfering whenever their interests demanded it. The USSR, unhampered at first by such interests, was able to give an impression of the disinterested benefactor and enjoyed considerable popularity and freedom of action.

The Soviet government soon offered development aid, armaments and the services of technical and military advisers apparently free of strings. While western countries were seen to be supporting mainly pro-western Arab governments – one reason why Nasser found it difficult to acquire the arms he wanted from the western powers – the Soviets did not make such political conditions. They took care not to insist on ideological conformity on the part of their Arab clients; for a time it was possible for an Arab leader to put all important communists in his country behind bars and yet qualify for Soviet economic and military aid. The Soviets' aim was to combat the West in the Mediterranean area and they were ready to aid any country that was engaged in the struggle against 'colonialism and imperialism'. But the main Arab effort was directed against Israel and this too was receiving Soviet support. The Soviets had joined the chorus of anti-Israel propaganda, branding the Jewish State as an outpost of western imperialism. Yet there was never any concrete evidence that the Soviet Union, like the Arab states, sought the destruction of Israel. Rather one may assume that the USSR wanted to use the Arab states to gain influence and power in the Middle East at the expense of the West.

The western powers had great difficulty in countering this Soviet involvement. Because of their physical presence and their wish to defend their interests, unconditional aid to those who sought it was not always a possible approach for them. Neither their economic or military aid, nor their alliance politics were effective weapons against the intrusion of the USSR into their traditional sphere of interest. To the expressed Soviet aim of 'anti-imperialism', which found a sympathetic echo in many Arab hearts, the West was able to oppose only its own doctrine of 'anti-communism', which, if it impressed the Arab right-wing governments, was treated with indifference elsewhere. The Soviets appeared to be supporting, somewhat paradoxically, non-alignment as well as Arab independence, which was what the Arabs wanted. The Americans countered with the Eisenhower Doctrine, by which they asserted the need to protect American interests 'against the dangers of international communism' and the right to use the US fighting forces if necessary against aggression by a communist controlled state. But to the Arabs it was 'not a struggle between western democracy and communism, but against western tutelage'.[3]

The situation created by the new Soviet role in the Middle East increased the dilemma of the United States. The arms deliveries from eastern Europe to Israel's most powerful antagonist increased the threat to Israeli security and aggravated the already serious instability in the Middle East. With all the undoubted concern of the United States for the survival and security of Israel, the Americans could not afford to be seen to give one-sided arms support to the Israelis if they wanted to preserve their influence in the Arab world, even with those Arab countries like Saudi Arabia, Jordan and Iraq which for internal reasons remained supporters of the West. Nor did the Americans wish Nasser's Egypt, which continued openly to profess non-alignment, to slip further into the Soviet orbit. That undoubtedly explains Washington's strong opposition to the military intervention launched by the British, French and Israelis, for reasons of their own, in the Suez Canal area in 1956. Meanwhile, the new great power rivalry gave a golden opportunity to President Nasser and other Arab rulers to play off the two superpowers against each other, in this way obtaining economic aid and political concessions from both sides. But the Soviet intervention in the Middle East was not the only feature in superpower relations at the time: the main theatre of the Cold War was Europe.

EUROPE

While tension was rising in the Middle East during the mid-1950s, important political developments were also taking place in Europe. After Stalin's death in the spring of 1953 Soviet foreign policy was for some time uncertain. The need for a reappraisal seemed to present itself not only because the change of leadership brought with it new political attitudes, but because at the end of the Stalin era Soviet foreign policy in Europe had gone through one of the most severely testing periods since World War Two. The outcome of this was a substantial defeat for the USSR. In the early 1950s the main diplomatic thrust of the Soviets was directed against what they perceived as the twin threat of west European unification and the rearmament of the Federal Republic. Judging by Soviet statements of the period, Stalin, in order to prevent both, had been prepared to pay a high political price: nothing less than the abandonment of Soviet hegemony over the German Democratic Republic in favour of the establishment of a reunited, lightly armed and neutral Germany. This offer, made by Stalin in the last year of his life, was rejected by the West, including the Federal Government, where preparations for west European integration and a treaty allowing the Federal Republic to re-establish fighting forces continued apace. In October 1954 the western powers and the Federal Republic signed the Paris Agreements ending the occupation of West Germany and allowing it to rearm and to join NATO. Apart from minor Allied reserve powers which remained, the Federal Republic was now a sovereign state, free to conduct its own foreign policy. But to the Soviets West German rearmament constituted a new threat to the USSR's security, the more so since this new military power was to be firmly anchored in the military and strategic system of the West. The Soviet government's reaction to these events was therefore to abandon all thoughts of German reunification and instead to throw all its weight behind the preservation of the status quo. The USSR now formed the Warsaw Pact alliance system into which the German Democratic Republic was incorporated, thus binding it more closely still to the Soviet Bloc.

Unlike the Soviets, the western powers, particularly the Federal Republic, were not prepared to abandon attempts to reunify Germany. Chancellor Adenauer hoped that by a combination of his own country's rearmament, European unification and the general strengthening of the West the USSR could one day be pressed – by peaceful, diplomatic means, it was understood – to abandon its hold on the German Democratic Republic, thereby making it possible to create a free, reunified German state based on western-style democratic principles and sufficiently strong to preserve its independence. In this he was assured of western support. To make this point the Federal Government established a claim that it, being the only freely elected German government, could express the political will of the whole German people, that it was indeed the only German government that was entitled to speak for all Germans. The implication of this was that the government of the German Democratic Republic, undemocratic and oppressed by a foreign power, the USSR, had no legal status in the international community and no right in international law to represent its own population. This West German claim of 'sole representation' was interpreted in practice to mean that neither the Federal Republic nor any other state should accord recognition to the German Democratic Republic and that all negotiations relating to Germany or any German question should be conducted through the West German government. To give this claim teeth, the Federal Government in 1955 established the principle – which became known as the Hallstein Doctrine – that if any government that entertained friendly relations with the Federal Republic contravened the claim by recognising the German Democratic Republic or establishing diplomatic relations with it, that would be regarded as an unfriendly act vis-à-vis the Federal Republic. In practice this meant that the Federal Republic would break off diplomatic and possibly economic relations with the offending country. It was argued that international recognition of both German states, leading to the establishment of two German diplomatic missions in the capitals of many

22

states, would gradually erode the concept of 'one Germany' and render Bonn's reunification policy ineffective. The Claim of Sole Representation received the support of the western powers and was written into the Paris Agreements. Together with the Hallstein Doctrine, it became an important instrument of West German foreign policy until the end of the 1960s. Both were aimed principally at the non-aligned countries since they rather than the Communist Bloc or western states might be tempted to establish relations with both Germanies.

The Soviet government, having proclaimed its insistence on the status quo, meaning the continuation of the two separate German states, now proceeded to recognise the Federal Republic and invited the Federal Government and other governments all over the world to recognise the German Democratic Republic as an independent sovereign state. So the East–West conflict over Germany shifted to a new level: for the next 15 years it became the battle over the recognition and the international status of the German Democratic Republic. International recognition of the East German state in fact soon became one of the most vigorously pursued aims of the Soviet and East German governments, while efforts to prevent this happening were now the cornerstone of the foreign policy of the Federal Government in Bonn. This particular aspect of the Cold War was fought out vigorously in the diplomatic missions of the Third World, that is, in countries which professed to be neutral. Both sides in the Cold War offered generous development aid in attempts to win these countries over to their point of view. As the Third World includes the Arab states, most of which profess to be non-aligned, it is here that the battle over the recognition of the German Democratic Republic impinged on West Germany's relations with Israel, as long as the latter was locked in a fierce conflict with its Arab neighbours.

The Paris Agreements, which came into force in May 1955, were a triumph for Adenauer's government in Bonn in that they returned the West German state to sovereignty. But the newly won independence and relative freedom of action in the international community exacted a price: it plunged the Federal Republic deeply into the Cold War. The new situation in Europe was bound to deepen the East–West conflict and to further embitter the relations between Bonn and Moscow and even more between the two German states. Adenauer's 'policy of strength' as it came to be called, by which he hoped ultimately to wrench the German Democratic Republic from the grip of the Soviet Union, depended on increased western rather than greater West German power. This made the Federal Republic more dependent on the goodwill of the USA. As for the Hallstein Doctrine, it effectively prevented for a time Third World countries, in dire need of economic aid, from recognising the German

Democratic Republic. Its danger was that it could be used as an instrument of political blackmail: if a country were one day to ignore the cutting of economic aid and even threaten to recognise the German Democratic Republic, it could thereby force a change on the policy of the Federal Government. It is clear, for example, that with the extension of the Cold War to the eastern Mediterranean, the Federal Republic, burdened with the Hallstein Doctrine, could not indefinitely disregard the Arabs' reactions to its pro-Israel policy.

GERMANY AND THE ARABS

The ambivalence of Arab–German relations in the last five decades is puzzling to the westerner and difficult to understand rationally. They have been shaped partly by the phenomenon of Germany's peculiar past relationship with western 'colonialism and imperialism', but inevitably also by the fact that the Federal Republic is a western state, a member of the Atlantic Pact and a friend of the United States. Both Germans and Arabs have often referred to the 'traditional Arab–German friendship'. The reasons for this are complex. For the Germans there was the romantic European conception in the nineteenth and early twentieth centuries of the ancient Arab culture; for the Arabs their interest in the success of German unification in the last century, given their own aspirations to unify their people; and, more important, their admiration for German efficiency and technological competence, as well as for the success, in two world wars, of German arms. Rather more realistically, some of the reasons for the Arabs' friendship towards the Germans could be found in recent times: for not only were the Germans blameless of any colonialist activities in the Middle East, they indeed fought against the European colonial powers in two world wars. If some Arabs may have had reservations about Hitler's methods and extermination policies, to most of them German politics were too remote, the conflict between western democracy and fascist totalitarianism too little understood for this to affect their loyalties. By contrast, their hatred of the Jews, whom the Germans had persecuted and who were now settling in Palestine, aided in the Arab view by the western imperialists, overshadowed all other considerations. And so there developed a fellow feeling from the idea that Germans and Arabs were together fighting the same enemy: colonialism; imperialism; Zionism.

In the First World War the Arabs had supported the western European powers against the Ottoman Empire and its ally, Germany. The reason was that Great Britain had made promises – as has been indicated – which led the Arabs to believe that the western Europeans would help them to

24

achieve their ideal: freedom from Turkish domination and national independence and unity. By the Second World War the British and French were considered to have betrayed the Arab cause. On the other hand Hitler had made important foreign policy gains in the Middle East at the expense of the western democracies, whose prestige among many non-western nations was consequently declining.[4] From 1940 the Axis powers were promoting an active anti-British campaign in Middle Eastern countries in order to create unrest with a view to damaging the British war effort. At that time the attitude of Arab politicians to the West was mostly hostile, at best neutral. The Axis effort therefore did not fall on deaf ears. Rommel's victories in North Africa gave encouragement to the Arab nationalists and raised their hopes that British power in the Mediterranean area would be broken. In fact British power was not seriously affected, having met with only minor challenges from the Arabs whose governments on the whole co-operated.

Arab sympathy for the Germans, however, survived Germany's defeat, humiliation and division. So did the notion that the Germans remained potential allies of the Arabs in their struggle for independence and unity. Thus Arab states were among the first to establish diplomatic relations with the Federal Republic when this became possible in 1951 following the 'Little Revision of the Occupation Statute' by the western occupying powers, signed after the Schuman Plan negotiations. Notwithstanding the occupation of West Germany by the western powers and its close political co-operation with them, the impression persisted in the Arab world that the Germans would not play the western game vis-à-vis the emerging Third World countries. The Federal Republic at the time was exempted from the general resentment felt by the Arabs towards the West. An Egyptian journalist wrote

> Now there is today a single western great power in the world which is still met with genuine trust throughout the whole Middle East ... and that is Germany ... a fact of which anyone in Islamic countries can convince himself.[5]

But this, according to many Arabs, imposed an obligation on the Federal Government 'not to disappoint the hopes of the Arab and Islamic world for essential aid and support in the struggle for the security of their existence'.[6] That the Federal Government in the early 1950s appeared ready for large restitution payments to Israel, to which the Arabs were vigorously opposed, was a cause for irritation, as was West German adherence to the European Coal and Steel Community, sponsored by a 'colonialist' power, France, and later to the Atlantic Alliance, led by 'imperialist' America. But even then many Arabs thought there were extenuating circumstances: that the Restitution Agreement with Israel was forced on the Federal Government by the western imperialists and entered into against the real wishes and interests of the German people. A similar view was taken in regard to the Federal Republic's membership of NATO and its pro-western policies generally.

This strange misinterpretation of the state of affairs may well be the reason why Arab governments continued to display an attitude of good-will towards the Federal Republic during the negotiations about the Restitution Agreement. Their disbelief that the Federal Government would willingly implement the agreement seems to have continued almost up to the actual point of signing. It would be difficult to see otherwise why strong pressure was not brought to bear by the Arabs on the Federal Government from the day Adenauer declared his willingness to meet the Israelis' demand for restitution. It must be remembered, of course, that disbelief over West German intentions was shared by others, not least by the Israelis. The reasons for this were doubts about the Germans' ability to meet their obligations in view of their precarious economic situation.

The Arabs' beliefs regarding the West Germans ignored the complete change that had taken place in the German political position since 1945 as a result of total defeat, occupation and above all the division of the world into two power blocs. The Federal Republic of the 1950s was in no way comparable to the German Reich of pre-war days. It nurtured no claim to world power. It had embarked on the difficult road back to respectability and equality among the nations. Adenauer and other West German politicians therefore had no hesitation in declaring to the Arab League and other Arab statesmen who tried to prevent it that the Luxembourg Agreement would be ratified as quickly as possible.

All the same, the chorus of Arab protests that began once agreement over German restitution had been reached with Israel caused concern in West German political circles and in the economy. Accusations made in Arab circles that by giving Israel massive financial support the Federal Republic was taking sides in the Arab–Israeli conflict were a sensitive issue for the West Germans, for it was unthinkable that a German government should become involved in any theatre of conflict so soon after the debacle of the Second World War. To the Arabs German resti-tution payments also appeared to neutralise the Arab boycott against Israel which was extended to firms of many nationalities trading with the Israelis. The Arab League now threatened to widen this boycott to affect West German firms supplying the Jewish State with goods under the Restitution Agreement. In view of the difficult problems the West German economy was still facing in the early 1950s, the country could ill afford the loss of useful export markets that had been painstakingly built up.

Representatives of West German industry and commerce now voiced their concern for their export markets and proposed that implementation of the Agreement be postponed. German export opportunities in the Middle East were good and likely to improve. Now the Arab League was even threatening to cut all economic ties with Bonn.

But the government stood firm. When an Arab delegation in Bonn, having failed to pressurise the Federal Government, turned to individual West German parliamentarians and former Nazis it was politely asked to leave the country. The West German ambassador in Cairo informed the Arab League that the Federal Republic would stand by the Restitution Agreement and rejected any postponement of ratification. Adenauer commented in an interview with a journalist: 'It would be shameful if we wavered in our determination only because we are being threatened with economic disadvantages. There are higher things than good business . . .'.[7] The Federal Government rejected the charge that it was taking sides in the Arab–Israeli conflict, maintaining that the supply of non-military goods to combatant states did not constitute a breach of neutrality in international law. At the same time, however, the government made concrete offers of development aid to Arab countries and by March 1953 a West German study group was in Cairo to investigate the possibility of participating in the proposed Aswan Dam scheme. It now seemed that the main Arab threat was receding, though occasional sniping by some Arab governments continued and Arab–German relations remained precarious.

Arab disunity and lack of concerted policies were one reason why the Germans won the day over the Restitution Agreement. Others undoubtedly were the realisation by the Arabs that a rupture of economic relations with the Federal Republic would ultimately hurt them more than the West Germans. Furthermore, the Arabs had few friends in the early 1950s. Although the USSR had begun to pay lip-service to their cause in the United Nations, there was as yet no hint of an active Soviet involvement. For all their dislike of the West, Middle Eastern states were still heavily dependent on western countries and of all these the Federal Republic was still the most acceptable one. As long as the West Germans moved no closer to Israel the Arabs would not or could not bring themselves to break with them.

So the storm had blown over for the time being. The Federal Republic did its utmost, while continuing to carry out its restitution obligations vis-à-vis Israel, to avoid any steps that could impair the now restored relations with the Arab world. Economic relations became increasingly important, both to the Arab states and the Federal Republic. But from 1955 the East–West conflict and the Federal Government's policy towards

eastern Europe began to create new complications. Henceforth the
Hallstein Doctrine was to dominate West German relations with both the
Arab states and Israel.

NOTES

1. Cf. Yair Evron: *The Middle East: Nations, Superpowers and Wars*, London: Elek, 1973, p. 192.
2. Cf. Arnold Toynbee, Letter to Jacob L. Talmon, July 1967 in *Encounter*, Vol. XXIX, No. 4, p. 68.
3. Bernard Lewis: 'Middle Eastern Reactions to Soviet Pressures', in *Middle East Journal*, Washington, DC, Spring 1956.
4. Mohammad Abediseid: *Die deutsch-arabischen Beziehungen*, Stuttgart, 1976, pp. 28–9.
5. Abdel Megid Amin: 'Deutsche Orient Politik heute' in *Aussenpolitik*, Jan. 1954, pp. 27–36.
6. Ibid.
7. Interview with Ernst Friedlander: Kreysler Jungfer, p. 39 (quoted in Deutschkron, op. cit., p. 91).

3

The Question of Diplomatic Relations

Israel's continued isolation, resulting from the hostility of its Arab neighbours, who were increasingly supported by the Communist Bloc both
diplomatically and militarily, made further shifts in the foreign policy of
the Jewish State an urgent necessity. With the United States cool because
of Israel's attack on Egypt during the Suez crisis, Israel's search for
friends inevitably led its government towards Bonn. Just at the moment,
however, when the Israeli government had overcome its inhibitions
towards the Germans and declared its willingness to establish normal
diplomatic relations with the Federal Republic, the West German government held back. Not that there was any evident change of heart in the
Federal Republic regarding Israel or the West Germans' acknowledge-
ment of the debt they owed to the Jewish – and by implication the Israeli
– people. Judging by the statements made by government ministers and
spokesmen, the wish to normalise West Germany's relations with Israel
was not in doubt and would be realised 'when the time was ripe'. But the
Federal Republic, since its return to sovereignty, had begun to play a
modest role in international politics, not only in pursuit of its own vital
interest, the reunification of Germany, but also as a kind of rich, beneficent
uncle to the Arab world, where it would contribute to preserving western
influence and keeping Soviet influence at bay. Both required the Federal
Republic to maintain a good relationship with the Arab states, which in
turn restricted its relations with Israel.

The heightening of tension in the Middle East in the course of 1955
brought increasing danger to the State of Israel. The rivalries and conflicts
within the Arab world caused by Iraq joining the Baghdad Pact, rather
than benefiting Israel, increased the threats along its borders. For the
Arab statesman most angered by Iraq's defection from Arab League
policies was President Nasser of Egypt who was increasingly championing
the cause of Arab unity. To him the Baghdad Pact was a moral defeat and
a setback to his own leadership of the Arab world in the struggle for
independence from western domination. Nasser's frustration manifested

itself in greater hostility towards Israel. The number of incidents provoked by Egypt on its border with Israel increased, as did the incursions into Israeli territory by the Fedayeen sabotage and other terrorist missions. This new militancy on the part of the Egyptian leader was designed to keep the pot boiling in the Middle East, but primarily also to reassert his leadership among the Arab states by pursuing the anti-Israel cause with greater vigour. The Israelis for their part had additional worries. They too were nervous about the possible consequences of the Baghdad Pact. If its aim was the defence of the Middle East against the spread of Soviet influence in the area, the arms supplied for this purpose by the West to Israel's neighbours could be used against Israel instead. To this was added, six months later, the probability that Soviet arms would soon be sent in large quantities to Israel's most powerful and militant enemy, Egypt. All in all these events raised in the minds of Israeli politicians and military men the thought that a preventive strike against one or several of Israel's neighbours might bring advantages. There was thus a danger that either side might start a war.

The western powers were anxious to prevent an outbreak of hostilities in the Middle East, one consequence of which could have been that a war would have strengthened Soviet influence at their expense. But they were in a difficult position. The United Kingdom was engaged in the Middle East as a member of the Baghdad Pact, which it would have liked to see enlarged by the inclusion of more Arab states. The United States, though committed to the preservation of the integrity of Israel, needed a measure of Arab goodwill in order to keep the Soviet Union out of the area. This is why they were anxious that the West Germans too should cultivate their good relations with the Arab states. The Federal Republic was already developing its economic ties with the Arabs. A serious dilemma was now facing firstly the United States because they considered Israel as the only reliable western state in the Middle East and secondly the Federal Republic, because it was bound to Israel for reasons of conscience. Only the French seemed less concerned about Arab friendship at that time. Their chief interest was to maintain their position in their North African Arab dependencies in some of which revolts were threatening.

The Israeli government also faced difficulties. Israel had had to move away from non-alignment because of increasing Soviet hostility and economic dependence on the United States, without, however, participating in any western alliance politics. Internally there was no consensus as to the precise direction of Israeli foreign policy: the extreme left-wing opposition parties were exerting pressure on the government of Ben-Gurion to move back towards neutrality and to avoid antagonising the USSR. These views at times spilled over into the less left-wing parties supporting the government. In the face of the new threats from some of its Arab neighbours, and especially when they began to receive weapons from the Soviet Bloc, Israel had to secure its own sources of armaments. The only large-scale suppliers at that time were the French. But this source was regarded as insufficient and unreliable because of the instability of French governments under the Fourth Republic.[1] There remained the Federal Republic but that presented a serious obstacle: arms deals with the Germans would have been difficult for the Israeli public to accept.

The Israeli government could not afford to be irrational. Not only did the country's insecure situation demand that it accepted the hand of friendship whenever it was held out: the Federal Republic was becoming more important both in economic terms and as a factor in world politics. The West German economy had by the mid-1950s become one of the strongest in Europe. West Germany had become a sovereign state, had joined the North Atlantic Alliance and was building up a new armaments industry. In Europe it had settled its age-long feuds with its neighbour, France, with which it was becoming closely associated. It was playing a major part in efforts to unite Europe. As an ally of the United States and France, Israel's two most trusted friends, the Federal Republic could not be ignored. Above all the Federal Government had, so it seemed to Israel, accepted a measure of responsibility for the injuries inflicted on the Jewish people by the Third Reich and made a commitment to the wellbeing of the Jewish State. Finally, the Federal Republic was increasing its trade links with the Arab states and actively cultivating good political relations with them. In theory at least the possibility that the West Germans might one day find it more advantageous to move closer to the Arabs and cool their attitude towards Israel could not be entirely excluded.

The question that was to dominate German–Israeli politics from 1955 onwards was that of diplomatic relations. Because of the attitude of the Israeli public to the Germans an exchange of diplomatic missions had not been possible. As one Israeli commentator expressed it:

> In the first years of the existence of the two states the mere thought of it would have seemed absurd to any Jew, in view of the events of the recent past, since even the conclusion of the Luxembourg Agreement had provoked endless discussions and passionate rejection among the Israeli population.[2]

For the Germans, on the other hand, full diplomatic relations with Israel in the early 1950s would have been highly desirable; to be able to establish such relations would have set the seal to German–Jewish reconciliation, initiated, so it was hoped in Germany, by the Luxembourg Agreement, and would have constituted a further step towards Germany's rehabilitation in the eyes of the world. During the debates on the ratification of the Agreement in the Bundestag Adenauer said:

> We have just hopes that the conclusion of this treaty may lead to a completely new relationship between the German and the Jewish people as well as to the normalisation of relations between the Federal Republic and the State of Israel.[3]

But official Germany was prepared to exercise patience: the healing process would need time and the Federal Government would not press the issue, but would wait until Israel was ready.

In practice the minimum machinery for carrying on the daily business, such as there was, between the two governments did exist. At the conclusion of the Restitution Agreement the Israel Mission, without diplomatic status but whose head was given the title of ambassador, was established significantly not in Bonn but in nearby Cologne. Its sole purpose at first was to purchase goods for Israel under the Agreement and supervise its smooth running. As, however, travel between the two countries became possible and an increasing number of Germans wanted to visit Israel, the Cologne Mission took on the function of issuing visas to aspiring visitors, while Israeli visitors to the Federal Republic – their number was at first severely restricted by the Israeli government – were similarly served by the British consulate in Haifa acting for the West German government. Technically therefore there seemed no urgency for changing the situation: the Restitution Agreement was working smoothly, trade between the two countries outside the Agreement was increasing, political relations were slowly improving. In practice everything could be handled by the existing machinery or by government emissaries even though there was no equivalent German mission in Israel. The main snag was the establishment of personal contacts and the obtaining of information by each government on the situation and political trends in the other country. For this the Cologne Mission, having been endowed with largely economic functions, was ill-equipped and lacked full diplomatic status. The Federal Government for example had to rely on a major West German news agency and its representative in Tel Aviv.

The political changes and pressures of 1955 gradually brought the Israeli government round to the view that progress in the question of diplomatic relations with the Federal Republic was a matter of urgency. There is some indication that a recognition of the importance of the West German state and the need ultimately to come to terms with it existed among Israeli politicians from about the time the Restitution Agreement came into force. By 1955 leaders of world Jewish organisations were also pressing the Israeli government to establish diplomatic relations with the Federal Republic. Throughout the year rumours were circulating from time to time in the West German and Israeli press that diplomatic relations were about to be established, only to be denied by one side or the other. But 1955 was an election year in Israel, and the government feared that the emotions which such a step would release might be exploited by the opposition parties. By October of that year, the Prime Minister designate, Moshe Sharett, in an informal press interview, went no further than holding out the possiblity of a West German consulate in Tel Aviv. He excluded full diplomatic relations for the time being.[4]

The change seems to have come quite suddenly early in 1956. On 14 March of that year the *Frankfurter Rundschau* reported a statement by the Israeli government that it would no longer object to the establishment of full diplomatic relations. It seemed that at last the government, well aware now of the need for such a step, had found the strength to accede to what had long been a desire of the West German government. Was it that Israeli politicians were conscious of developments within the Federal Government and realised suddenly that time was running out? On the same 14 March the head of the Israeli mission in Cologne received a letter from the Federal Foreign Minister, von Brentano, informing him that 'the Federal Government agrees in principle to establishing an agency in Israel which would correspond approximately to the Israeli mission in Cologne'.[5] This statement was vague enough. It certainly offered the Israelis less than their government had now, according to their official statement, agreed to. For the Cologne mission did not enjoy diplomatic status and a corresponding office in Tel Aviv would in no way have meant full diplomatic relations. The minister's letter simply returned to an idea vaguely canvassed by the Israelis during the previous year but now considerably modified. It nevertheless makes the point that the West German mission could be the forerunner to full diplomatic relations in the future. A few days after the receipt of the letter the FAZ quoted 'competent circles' in Bonn as denying reports that diplomatic relations with Israel were imminent. The matter had been discussed, the paper writes, but both sides had agreed that the date was still open and the matter not at present under consideration.[6] It was back to square one.

The change of mind by the Federal Government must be related to the main events that occurred in Europe in 1955 and above all to the West German reactions to these. The Hallstein Doctrine, the Federal Government's counter-offensive against the Soviet insistence on a permanent division of Germany, had first been formulated at the end of 1955 at a conference of West German ambassadors called home to Bonn to receive instructions on the new lines of West German foreign policy. In the spring of 1956 it was to be put to the test for the first time. Rumours of the impending establishment of diplomatic relations between the Federal Republic and Israel, circulating in the early months of 1956 and culminating in reports that Israel and the Federal Government had agreed, caused strong reactions in the Arab states. The Middle East Information Agency reported that the Council of the Arab League was about to consider its members' joint attitude to the Federal Republic in the event of its recognising Israel.[7] In an interview with INS correspondent Karl von Wiegand Nasser was reported to have said: 'We have so far refrained from recognising the German Democratic Republic and establishing diplomatic relations with it', adding, ironically, 'recognition of Israel by the Federal Republic could remove our inhibitions, for it would be a hostile act vis-à-vis the Arab nations'.[8] This turning round of the Hallstein Doctrine against Bonn by the Egyptian leader was soon followed by a chorus of protests from Arab politicians directed now not only at the reported plans for diplomatic relations between West Germany and Israel but also once more against the Restitution Agreement. It seemed that in the threat to recognise the German Democratic Republic the Arabs had found a weapon against the Federal Government not only against its recognising Israel but also suitable to stop the flow of goods and services from West Germany which were helping to boost the Israeli economy and, in the Arab view, increasing Israel's war potential. The Federal Government, concerned about the feedback it was obtaining from its ambassadors in the Arab capitals, summoned them in May 1956 to a conference in Istanbul chaired by Walter Hallstein, then State Secretary in the Foreign Office. He seems to have received the clear impression from them that the policy of the Federal Government towards the GDR and the West German claim to speak for all Germans would be at risk if Bonn established diplomatic relations with Israel. There was imminent danger that some Arab states would recognise the GDR. This would necessitate the rupture by the Federal Government of diplomatic relations with such countries, with the probability that it would endanger the increasingly lucrative trade with these countries, as well as undermining Bonn's reunification policy. These were risks that the Federal Government was not willing to take.

So the warnings given by the ambassadors to Secretary of State Hallstein at the Istanbul conference seemed to have sufficed to deter the Federal Government from taking any steps in the direction of normalising its relations with Israel. It is not clear whether the proposed German mission in Tel Aviv also came under the axe because of Arab pressure or whether the Israelis, now ready for full diplomatic relations, were no longer willing to accept anything less. What is certain is that the idea of full official recognition of Israel had been shelved indefinitely. Thereafter, though rumours about the imminence of the establishment of diplomatic relations circulated frequently, nothing serious was done in the matter for the next nine years and the envisaged German mission in Israel was never set up.

The Federal Government from then on had some difficulty presenting its policy to the German public, a majority of which was strongly sympathetic to Israel and baulked at the illogicality of the government's attitude. Was it not at least surprising, people asked, that Israel of all countries, to which the German people owed a great moral debt, should be virtually the only country in the world that the Federal Government would exclude from the normal conventions governing international relations? The *Süddeutsche Zeitung* asked if the Federal Government could afford to be pressurised by other states. The establishment of diplomatic relations with Israel would be a normal act of sovereignty on the part of the Federal Republic.[9] Others thought that the Federal Government should have shown on this occasion the same steadfastness it showed earlier, for example after the signing of the Restitution Agreement, and ignored Arab warnings, hoping that the Arabs would not implement their threats. But the Federal Government's policy in respect of reunification was by now receiving priority over everything else, even the special relationship with Israel. This was not to say that all thoughts of this had been abandoned. The Federal Government continued to press for further improvement and to hold out hope for diplomatic relations 'when the time was ripe'.

With hindsight it is difficult to see how recognition of Israel by West Germany could ever be reconciled with the operation of the Hallstein Doctrine unless the Federal Government was prepared to take some risks. But there is no doubt that Arab recognition of the East Berlin regime, had their governments carried out their threats, would have dealt a serious blow to the prospects of German reunification as they were then perceived in the West. The Hallstein Doctrine, often criticised as inflexible and legalistic, made some sense as long as the Federal Government hoped to mobilise world diplomatic pressure, backed by a measure of military strength in the West, to oblige a reluctant Soviet Union to give up its control of the German Democratic Republic and permit the reunification of Germany. A universal recognition of two German states could well have consigned the whole question of the division of Germany to oblivion, at first in the Third World, where it did not play the important role it played in Europe, then in the United Nations, where the non-committed states because of their numbers wielded some influence, and ultimately perhaps in the West as well. For in the West, where not every government was happy about the prospect of a reunited Germany, the support for German reunification can be seen mainly as a tactic in the Cold War; and the Cold War made sense in the West only as long as there was a strong imbalance between the two contenders in favour of the West. This situation did not change until the 1960s. But even when it did, the Middle East remained an area of East–West conflict and to prevent the growth of Soviet influence there continued to be western policy. It was here that the Federal Republic had a role to play.

For the difficulties which the western powers were experiencing in holding their position in the eastern Mediterranean region combined with the Bonn Government's own policy to give the Federal Republic a unique chance. The West German state was not only now becoming attractive in Israeli eyes, it was also becoming increasingly desirable to the Arab rulers, despite the Federal Republic's membership of the western alliance and its restitution agreement with Israel. They obtained from it both large amounts of economic aid and, in general, sympathy for their development problems. With this and its past record of behaviour towards colonial countries it was able to present to the Arabs the only acceptable face among all the nations of the West. The West Germans had to consider how they could best exploit this asset both for themselves and for the benefit of the West as a whole. There was therefore more at stake than the widespread recognition of the GDR which would have occurred if Nasser had carried out his threat and other Arab leaders had followed his example. West Germany had to keep its relations with the Arab states at the best possible level by providing the development aid and technical know-how they wanted, thereby making it less attractive for them to risk the rigours of the Hallstein Doctrine, helping them at the same time to preserve their independence from both East and West. Had they taken a false step, leading ultimately to the application of the Hallstein Doctrine, this would have created an economic and political vacuum which could easily have been filled by the Soviet Bloc. Communist countries would have gained the opportunity to step in where the Federal Republic would have had to leave. Nasser's politics over the years, despite his anti-western propaganda and occasional leaning towards the Soviet Bloc, show this was not what he wanted. In this situation it is not surprising that the Federal Government obtained backing for its reluctance to recognise Israel from the government of the United States, despite the latter's sympathy for and support of the Jewish State.[10]

But while by this policy the West Germans ingratiated themselves with the Arabs, their dilemma increased. For the moral commitment to Israel remained. The Federal Government had stressed it to the public for years and it continued for some time to be a cornerstone of the country's internal and external rehabilitation.

An early example of the embarrassment caused to the West German government by such a conflict of loyalties was the Suez Crisis of October–November 1956. The trouble arose when President Nasser nationalised the Suez Canal in retaliation for the refusal by the United States government to implement its promise of finance for the Aswan Dam scheme. The Egyptians, who shortly before had shown willingness to co-operate with the West, had come under suspicion because of the arms contracts and increasing economic and military links with the Soviet Bloc. The bombshell of the nationalisation of the Suez Canal prompted the United Kingdom, together with France, to mount an attack on Port Said, the Egyptian port at the northern end of the Canal. At the same time Israel invaded Egypt on land through the Sinai Desert. The attack had been concerted by the three countries for different reasons: the French feared the spread of Arab nationalism into their North African protectorates and the supply of arms by Egypt to the Algerian rebels. The Israelis were concerned about the build-up of Soviet arms in and the growing number of incursions by Arab terrorists from Egypt.

Nasser's tour de force over the Suez Canal, which aroused much admiration in the Arab world, was a victory for Arab nationalism and heralded the end of British and French influence in the eastern Mediterranean. It was not in itself a victory for Soviet penetration, but the psychological impact of the military coup, staged as it was by advanced European nations against a country of the Third World which still professed non-alignment was likely to inflict severe damage on the western position in the area. The Americans clearly understood this and forced the premature abandonment of the venture, leaving the three participants with a severe loss of prestige and without having attained any of their objectives.

To the West German government the Middle Eastern crisis was a matter of deep concern. The attack on Suez, in which Israel had participated, brought a wave of criticism in the German press and public opinion of the western European allies, Great Britain and France, as well as of the Israelis. The Adenauer government now had to face the possibility of conflicting claims by both Israel and Egypt for support. Even more serious was the rift in the western alliance system resulting from the American diplomatic intervention against Great Britain and France. West German aspirations for reunification required a strong and united West and the future development of the 'policy of strength' was thought to be in danger. There was a conflict between the Federal Republic's new role in the Middle East and its increasing economic interests in the Arab states on the one hand and its close relations with France, on which its policy of European unity was based, on the other. But most dangerous of all was the joining of forces of the governments of the two superpowers to put an end to the attack on Suez. Concertation of policies by the United States and the USSR was a nightmare that was to haunt western Europe but particularly the Federal Republic during the late 1950s and 1960s. 'It remains a milestone of post-war politics' says Adenauer in his memoirs, 'that in the Suez Canal Crisis the United States of America and the Soviet Union voted for the first time together in the United Nations against two states of NATO.'[11]

Adenauer is also sharply critical of the American action of withdrawing financial support for the Aswan Dam project, which clearly was of vital importance to Egypt and its growing population. He reproaches the US government for being badly informed about the situation in the Middle East and in Egypt in particular. That the Soviets were at that time tempting Nasser with generous offers must have been another reason for concern to Adenauer, who was as anxious as other western leaders that western influence in the Middle East should not be replaced by influence of the Soviet Bloc.

Again Adenauer's government was critical of the Anglo-French attack on Egypt. In this case, however, his criticism was tempered by his anxiety over progress with European integration. Negotiations about this were at that time directed towards the creation of the European Economic Community and had reached an advanced stage. The Suez crisis and the serious rift within the Western Alliance had in Adenauer's view reinforced the need for European unity, especially in the field of foreign policy. The Federal Republic's increasing commercial interests in the Arab world and its incipient role as the 'honest man' of the West forbade any strong sympathies with the western European countries involved in the military action over the Suez Canal, while on the other hand they were allies in NATO and one was a partner in the prospective European Communities. And the United States remained the leader of the Atlantic Alliance on which West German security and the future of a united Germany depended. In this extraordinary, complex situation facing the West German government it had little alternative but to profess complete neutrality in the Middle Eastern conflict, at the same time to do its best to help defuse the situation and bring about a return to normality as speedily as possible. As far as Israel was concerned, the West German attitude was unambiguous: complete neutrality in the Israeli–Egyptian conflict, but no change in its policy towards Israel, which had remained consistent ever since the signing of the Luxembourg Agreement. The Federal Government did not renege on this treaty. When therefore the possibility arose that the United Nations might impose economic sanctions on Israel because it did not immediately withdraw from Egyptian territory, the Federal Government announced that it would not participate. It justified this attitude by stating that the Federal Republic was not a member of the UN and therefore not under any legal obligation to carry out its sanctions policy.

The West German government's stance of neutrality, though in a sense negative and falling short of more active support that both sides may have hoped for, nevertheless seems to have caused satisfaction to all parties. The Israelis, who at first had doubted the German willingness to implement the Luxembourg Agreement, showed appreciation at the Federal Government's assurance that restitution payments would continue even if the UN applied economic sanctions. The Arab states, while continuing to be generally dissatisfied over the Luxembourg Agreement, seemed satisfied about the Federal Government's critical attitude, despite its many ties with Western Europe and the United States, towards the British, French and Israelis over the attack on Suez and its continued manifestation of goodwill towards the Arabs. For the west European nations there was no question of a wider breach in the Atlantic Alliance: all were anxious to heal the malaise as quickly as possible. Neither was there any doubt about the need for continuing progress in the integration of western Europe.

If the Adenauer government's neutrality policy had paid off as far as its search for good relations with all Middle Eastern countries was concerned, this was not to be for long. In 1957, as the dust of the Suez Crisis gradually settled, the question of diplomatic relations between the Federal Republic and Israel arose again and caused strong reactions on the Arab side. There were many reports circulating in Israel, the Federal Republic and the United States in the spring and early summer of 1957 that the question of diplomatic relations was being examined anew by both Israel and West Germany. Some said that a decision was imminent. There seems little doubt that the West German government was testing the temperature both as far as the Arabs and the western allies were concerned. The Arab reaction was not surprisingly one of unmitigated hostility, coupled with threats to recognise the GDR. Undoubtedly because of this the American reaction was cool. The Israelis were angered by their failure to enlist American diplomatic support while the West Germans felt embarrassed by threats from the Arab League office in Bonn. But the political risks were too great and clearly the United States government agreed that the moment for diplomatic relations was not opportune.

At about this time the Israelis began discreetly to apply pressure on Bonn though it never went as far as an official request. In their view the

Arab threats were a bluff, an opinion they expressed repeatedly right up to the time when Israel and West Germany finally normalised their relations. After all, the Arabs had used the barrage of threats to recognise East Berlin five years ago in order to oblige Bonn to abandon the Restitution Agreement. They had never carried out their threat and would find good reasons for not doing so now. The matter was debated in the Knesset in mid-July 1957 when Ben-Gurion stated that the establishment of friendly and diplomatic relations with West Germany was desirable and in Israel's interest. This was the most categorical statement so far. The Knesset thereupon confirmed this by a vote of 41 to 16, the clearest expression of Israel's new attitude to Germany to date. The reason is not difficult to see: the Suez Campaign had not alleviated Israel's isolation, it had if anything aggravated it. The hostile Anglo-French attitude towards their 'ally' during the campaign was a disappointment and must have sown doubts in Israeli politicians' minds about the reliability of France, which had been Israel's staunchest friend and greatest supplier of arms.[12] The premature abandonment of the military operations, the diplomatic defeat and humiliation of the west European states were another, and the inevitable decline of their influence in the Middle East was not in Israel's interest. The stern attitude of the United States, their other friend and ally, and what amounted virtually to support for the USSR in the United Nations over the question of Israeli withdrawal from Sinai and the Gaza Strip were a further cause for anxiety. The current situation made it more essential than ever that Israel should exploit its developing special relationship with the Federal Republic, especially as there was increasing support and goodwill for Israel among the West Germans. *Süddeutsche Zeitung* commented as follows:

> The Federal Government declares its readiness in principle to establish diplomatic relations. It sounds weak. Evidently it is still worried about the Arabs. But the establishment of diplomatic relations with a member of UNO, recognised by all except the Arabs, was no justification for Arab counter-measures. Temporary upsets should be overcome by diplomatic skill. The Federal Government should not hesitate to heed the decision of the Israeli parliament.[13]

Although the Israeli government made it a principle never formally to take the initiative, it was made clear to the West Germans in speeches and through contacts with German political leaders throughout the late 1950s and early 1960s, that Israel was anxious to normalise relations. From the time that the Knesset had voted in favour, Israeli Prime Minister Ben-Gurion never tired of expressing the view that the Germany of the post-war era was not the same as the Third Reich and that it was right that Israel should move closer to it. He continued to express this view in the face of strong resistance from among some of his own cabinet colleagues as well as from a section of the public, who had remained adamantly anti-German despite a general mellowing of the Israelis' attitude towards the Federal Republic. Whether Ben-Gurion sincerely desired reconciliation with the Germans or was only making a virtue of necessity is a question that his biographers may one day be able to answer. What is not in doubt is the energy and shrewdness with which he pursued his quest for better relations between the two countries. 'Man dies with his guilt', he said in one interview with an Israeli newspaper. 'If a German father was evil and his sons are good men then I have nothing against the sons ...', and further on: 'the young generation is different ... the present rulers are not identical with Nazis'. The German people were a people like all the others. He himself would go to Germany if there was a political need.[14] In another interview he expressed himself in favour of diplomatic relations, saying that the Federal Republic was with Israel on the side of western democracy against totalitarianism. The Federal Republic had accepted responsibility for the Nazi crimes and was paying compensation. This restitution, made for reasons of conscience, was unique in history.[15]

In a special vacation session of the Knesset, called by the opposition, Ben-Gurion made a speech containing three points: Israel was interested in preserving the status quo in the Middle East as long as possible; Israel would strengthen its defences as much as possible; and Israel was interested in acquiring as many friends as possible in the world. In connection with the last point he mentioned some Afro-Asian countries, but also the Federal Republic.[16] That there was no adequate response from the West German side was regretted and the unwillingness of the Federal Government to establish diplomatic relations was regarded by some Israelis as tacit support of Arab policies against Israel, dictated by the fear that Egypt might recognise the German Democratic Republic and withdraw its orders from German firms.[17]

The year 1958 was a year of crises and changes in the Middle East and brought an increase in tension almost everywhere. Behind it was the increasing activity of the Soviet Union, both in the Middle East and in Europe, especially towards the end of the year. 1958 saw a revolution in Iraq which swept away the monarchy and with it the pro-western government of Nuri-es-Said. The new regime which emerged out of the chaos, led by General Kassem, had strong communist participation. It withdrew from the Baghdad Pact alliance and declared itself non-aligned, leaning nevertheless towards the USSR from which it received economic and military aid. Another factor was the increased activity of President

Nasser's Egypt, which had recovered from the worst effects of the Suez campaign. In February Egypt joined with Syria, where communist influence was increasing, to form the United Arab Republic. Very soon the remaining pro-western Arab states began to complain of interference by the UAR in their internal affairs. At the request of their governments the USA sent marines into Lebanon and the United Kingdom sent paratroopers to Jordan. A takeover by Nasser, whose popularity among the Arab peoples was still growing, was feared by western governments. It would have meant a dangerous encirclement of Israel. The Egyptian leader was, moreover, intensifying his anti-Israel propaganda. The Arab boycott of Israeli goods and firms trading with the Jewish State was also strengthened and its effect was increasing.

1958 was also a year of crisis in Europe. Soviet scientists had been making progress in nuclear and rocket technology and there was evidence that they had the ability to send intercontinental missiles with nuclear warheads across the Atlantic. This would ultimately nullify the then current NATO strategy of massive deterrence or retaliation under which the USA had been building up its nuclear capability and establishing bases in many parts of the world, including the Middle East, from which to launch a nuclear attack on the Soviet Union, should it attack any member of the North Atlantic alliance. This strategy would clearly be of no value if the Soviets were able to strike back in kind.

The progress made by them, though this did not reach the capacity of the USA, prompted the Soviet leader Khrushchev to make the best of it while he had the chance. His interest was to obtain international recognition for the GDR and establish once and for all the permanence of the two German states. By issuing the Berlin Ultimatum in November 1958 he hoped to force the West to negotiate with the German Democratic Republic over Berlin – which would have meant recognition of the East German state – or if they refused, to 'solve' the problem of the divided city by obliging the West to abandon West Berlin. The Berlin crisis, which lasted until 1962 and during which the Berlin Wall was built, is important in that it heralded a change in superpower relationship from Cold War to détente and this in turn – as will be discussed later – affected American attitudes to the West German policy of strength and ultimately the Federal Government's own policy towards the GDR.

The crisis distracted attention from Soviet activities in the Middle East whose aims there were similar: to extend Soviet influence and obtain recognition for the GDR, thus upgrading its international status. In October 1958 Moscow signed an agreement with Egypt to finance the first stage of the Aswan Dam project. The West Germans, who with the encouragement of Washington had made an offer to do so, were pipped at the post. There followed a complex pattern of Egyptian rebuffs to the Federal Republic, and in some cases to the Soviet Bloc, as well as acceptance of offers of economic aid from both. Although Nasser was fighting against western influence he was also wary of Soviet power and had at times to walk a tightrope between his need to receive substantial economic and military benefits and his refusal to accede to Soviet wishes. The occasional offence given to the Federal Government, for example, over Cairo's relations with the GDR, may be seen in that light rather than as a deliberate affront to the Federal Republic. In January 1959 the Egyptian government received Otto Grotewohl, the Prime Minister of the German Democratic Republic. This event could be seen as a sop to Moscow in view of Nasser's attacks on the communists both inside and outside Egypt, which was viewed with less tolerance by the Soviet government than it had been earlier. Grotewohl, the second in command in East Berlin, was the first high dignitary of the GDR to visit a non-aligned country in an official capacity and this could not be ignored by Bonn. At the end of the visit it transpired that the Soviet and Egyptian governments had agreed to establish an East German consulate-general in Cairo. The West German government protested to Egypt but did not take any action. The assurances given by Nasser to the West German ambassador that the proposed consulate would not imply diplomatic recognition of the GDR and that such recognition was not contemplated seem to have satisfied the Federal Government. In 1961 it was announced that a consulate-general of the GDR would be opened in Damascus, the capital of Syria and now the 'northern region' of the United Arab Republic. Strong language was used by Bonn over this further action, regarded as a breach of promise made by Nasser in 1959. The West German ambassador in Cairo was recalled to Bonn on extended leave. Once again the Egyptian leader made soothing noises and gave assurances that no recognition of the German Democratic Republic was intended. No West German action followed. There were certainly signs that Nasser did not want to break off relations with the West. Despite the Egyptian president's pinpricks, the Bonn government had at that time reason to be satisfied that Nasser wished to keep his lines open to both East and West. This, the desire to maintain its position in the Arab world, the lucrative export markets and the crisis atmosphere reigning in the West over Berlin may all have combined to impose restraint on the Federal Government over the consulate.

The establishment of these offices in fact amounted to very little but it did indicate that the Egyptian leader was unlikely to be sensitive to West German political demands for ever. The efficacy of the Hallstein Doctrine declined during the 1960s. To the Americans, who were increasingly preoccupied with superpower détente, the continuing cold war between the Federal Republic and the Soviet Bloc became a burden. The mildness with which they reacted to the building of the Berlin Wall shocked the Adenauer government but it was indicative of the change that had taken place in US foreign policy. Soviet advances in the nuclear field had made a confrontation between the superpowers highly dangerous for everyone. Although the western powers stood their ground over Berlin, it was clear that neither side was prepared to risk a nuclear war. More than that, the superpowers in the 1960s were looking for areas of agreement in order to diminish the risk of such a war. Yet while this was happening the Federal Government, standing by its claim of sole representation and the Hallstein Doctrine, had by contrast manoeuvred itself into a position of immobilism, where it was out of step with western policies and in danger of isolation.

Adenauer's policy towards eastern Europe was also becoming unpopular in the Federal Republic itself. The Hallstein Doctrine was rapidly losing credibility when it was seen that the gulf between the two German states was widening and western support was declining. The immediate reason for building the Berlin Wall was to stop the thousands of refugees leaving the GDR and passing through West Berlin to the Federal Republic. But the West German media and public had understood that behind the East Berlin government's action was its aim to seal off the East German state from the West. It was an act of defiance against Adenauer's policy of strength, by which he hoped to coerce the USSR into abandoning the GDR and thus to reunify Germany. This policy was now seen by many to have failed and hopes for political reunification were receding. Yet the estrangement between the two parts of the German people was growing as the slanging match between their governments continued. Meanwhile Bonn's dangerous tightrope policy in the Middle East was maintained and Israel's resentment and the German public's impatience with the stalemate in German–Israeli relations was increasing.

Against the background of official German double-talk, pressure by public opinion and many politicians in the Federal Republic to establish diplomatic relations with Israel continued and even increased. People questioned the government's persistence in withholding from Israel what it had granted to most other nations in the world, while public figures and the press never ceased to refer to Germany's past and the moral debt that Germans owed to the Jews and the State of Israel. There was certainly no lack of reminders of the past. At the end of 1959 a rash of anti-Semitic incidents occurred in Germany but also in other western countries. They were characterised by the daubing of swastikas on Jewish community buildings and the desecration of Jewish cemeteries. Their international ramifications caused bewilderment, but it was also considered a mitigating circumstance for the Germans, in that the Federal Republic was only one country among many to suffer from this ignominy. Adenauer rightly played down the incidents, which continued for a few weeks into 1960, attributing them to the work of a small minority of incorrigibles and asserting that Nazism had no roots in contemporary German society.[18] But there was a feeling of outrage among the public over these excesses which in the words of the Minister of the Interior, Gerhard Schröder, 'went against the general will finally to overcome the most bestial and inexcusable chapter of Nazi history by restitution, reconciliation and tolerance'.[19]

Episodes of this kind, like the later, much more important event relating to the persecution of the Jews in Hitler's time, the Eichmann trial of 1961, could not, as it turned out, affect the official relationship between the two countries; they could only highlight the paradoxical nature of them. The West German government's attitude remained unchanged, despite increased public impatience. Nor did Israeli pressure increase. The Israeli government reacted diplomatically: it sent notes, identical in content, to the governments of all countries where anti-Semitic incidents had occurred, including the Federal Republic, warning them of the dangers of anti-Semitism and the need to crush it. Despite demonstrations by extreme right- and left-wing parties in Israel, Ben-Gurion was determined to promote his country's improving relationship with the West Germans and his confidence in Adenauer's government was unimpaired.

On the West German parliamentary scene there was some division. The SPD was wholeheartedly for the establishment of diplomatic relations with Israel and its parliamentary deputies like Carlo Schmid, Herbert Wehner and Willy Brandt, together with their leader Erich Ollenhauer expressed themselves frequently in favour of normalisation, as did spokesmen for the Trade Union Federation. In the early 1960s the SPD had come to oppose the Hallstein Doctrine as an ineffective and dangerous tool of West German foreign policy. They now argued that moral considerations far outweighed the expediency of avoiding Arab recognition of the GDR. Their sympathy lay naturally with the socialist government of Israel and the argument that Germany might suffer economic losses if the Arab states reacted did not impress them. There was also support for diplomatic relations among Free Democrats and Christian Democrats, but here there were some notable exceptions. There were reasons for this; being members of the government parties many agreed with the government's policy. It would also be surprising if the strong economic lobby, concerned about the dangers to future trade with the Arab countries, and even the smaller much more ill-defined group of former Nazis did not make some impact. When the matter was discussed in the Bundestag Foreign Affairs Committee, which met in closed session, the government carried the day: a majority of the Committee showed itself in favour of not normalising relations with Israel at that moment.[20]

The Federal Government for its part never failed to assert that it agreed to the establishment of normal relations with Israel in principle though for many years it repeated that 'the time was not ripe'. Reasons for this were rarely given and if they percolated through to the public they were regarded as unacceptable. Giving way to Arab threats of reprisals in a matter of international usage would have been an admission that Arab states could exert undue influence over West German foreign policy. Such statements as were made by members of the Federal Government thus did not give the Hallstein Doctrine as a reason for the anomaly but pointed to the dangers in the Middle East. The exchange of envoys with Israel would be unwise, according to Brentano on 4 November 1957, as it would increase tensions in the Middle East and it was not the task of the Federal Republic to do so through diplomatic steps.[21] More surprising is Adenauer's statement made in an interview with the Israeli evening paper *Ma'ariv* that diplomatic relations were against Israel's interests as they would provide new arguments for Arab unity against the Jewish State.[22] He denied that West German fears that the Arabs might recognise the GDR were the reasons for his government's attitude. In an interview with *Le Monde* he elaborated: 'It would be harmful to the Jewish State in the sense that it would stir up again the old hatred against Israel. Certainly the Arabs must recognise that the Zionist State is a fact, but one must make this evolution easy for them.'[23] In the debate on the subject of West German–Israeli relations one could argue, and some did, that the opposite was true: the reticence of the West German government could be regarded by the Arabs as a tacit recognition of and support for their anti-Israel policy and could even encourage them to increase their preparations for war against Israel, thus increasing tension in the Middle East.[24]

It seems that neither side had the truth, for there is no evidence that Arab hostility towards Israel was affected by West German policies or would have changed had the Federal Government altered its official position regarding Israel. The real reason for the West German government's attitude was fear of Arab recognition of the German Democratic Republic, followed by the inevitable application of sanctions under the Hallstein Doctrine, which might in turn have led to the Federal Republic losing its special position as an agent of the West in the Arab world. Although the Israeli public felt insulted and the Israeli media often criticised this Macchiavellian style of West German politics, the Jerusalem government, though sometimes critical, showed understanding. It had some reason to do so. For behind the scenes German–Israeli relations were developing at a different level, in a way that was of great importance to the security if not the existence of the Jewish State.

NOTES

1. Only socialist governments, friendly to the socialist Israeli government, could be relied on.
2. Deutschkron, op. cit., p. 103.
3. Shinnar, op. cit., p. 109.
4. *Südd. Ztg.*, 19 Oct. 1955.
5. Shinnar, op. cit., p. 113.
6. *Frankfurter Allgemeine Zeitung*, 23 March 1956.
7. *Bourse Egyptienne*, 2 April 1956.
8. Reported in *Südd. Ztg.*, 4 April 1956.
9. *Südd. Ztg.*, 5 April 1956.
10. Cf. *New York Times* report of 8 July 1957 stating that Abba Eban, then Israeli ambassador to the US, asked Secretary of State Foster Dulles during the latter's meeting with Adenauer in Washington, to press for diplomatic relations with Israel. Dulles agreed but merely mentioned it not to Adenauer but to von Brentano the Federal Foreign Minister. This greatly annoyed the Israelis. Nothing came of it.
11. K. Adenauer, *Erinnerungen, 1955–59*, vol. 3, p. 226, Frankfurt: Fischer Bücherei, 1969.
12. Although it was then suspected, and is now known, that the Suez attack was concerted by the three countries, this was not admitted at the time. When Great Britain and France were forced to withdraw from Suez they sent an 'ultimatum' to Israel to withdraw its forces from Sinai.
13. *Südd. Ztg.*, 19 July 1957.
14. *Ha'aretz* quoted by *Die Welt*, 6 Nov. 1959.
15. *Yediot Acharonot*, quoted by *Südd. Ztg.*, 5 July 1959.
16. Cf. Y.M. Ben-Gavriel: 'Israel und der Status quo' in *Aussenpolitik*, Dec. 1958, pp. 786–7.
17. Ben-Gavriel, ibid.
18. Dpa report, 16 Jan. 1960 quoted by J. Kreysler and K. Jungfer: *Deutsche-Israel-Politik: Entwicklung oder politische Masche*, ch. 5.
19. *Bulletin*, 7 Jan. 1960, quoted by Kreysler and Jungfer, op. cit.
20. *Tagesspiegel*, 22 Jan. 1960.
21. J. Seelbach: *Die Aufnahme der diplomatischen Beziehungen zu Israel als Problem der deutschen Politik seit 1955*. Anton Hain, 1970, p. 92.
22. *FAZ*, 7 Oct. 1959.
23. *Le Monde*, 8 Oct. 1959.
24. Franz Böhm in the *Frankfurter Hefte*, vol. 20, no. 2, Sept. 1965, pp. 601–25.

4

The Sale of Arms and a Secret Arms Agreement

One way in which the United States government showed its displeasure to Israel for its participation in the Suez campaign was to cut off arms supplies to the Jewish State. The Americans, who had been opposed to the attack on Egypt by Great Britain, France and Israel because they feared it would further antagonise the Arabs against the West, were now bound to adopt a much more even-handed approach to both the Arabs and Israel. From their point of view this was necessary to prevent further Soviet infiltration into the Arab world. Apart from the global power struggle between the two superpowers, the supply of Middle Eastern oil to the developed western nations became an increasingly vital factor. The United States government's refusal to supply arms placed the Israeli government in a serious situation. Israel was heavily dependent on the outside world for war material, since the Suez campaign had done nothing to decrease the danger it faced from its Arab neighbours. Because of the explosive situation in the Middle East after the Suez crisis most western countries were not prepared to involve themselves in Middle Eastern affairs by supplying arms. On the other hand the Communist Bloc countries were continuing their massive arms deliveries to the Arab states, especially to Egypt, which had lost great quantities of war material in the Suez campaign that needed to be replaced. As was said earlier, only France was willing and able to send weapons to Israel, but this was not sufficient. For Ben-Gurion to exploit the only remaining source, the Federal Republic, was a bold step, fraught with great political danger to his coalition government. A military agreement with Germany was the last straw for his opponents and many of his supporters. At one stage the discussions led to the break-up of the government. But it re-formed and the military relationship between Israel and the Federal Republic developed. Despite its secrecy it strongly affected the underlying trend of West German–Israeli relations over the years. When the secret was finally revealed, it paradoxically led to the normalisation of official relations between the two states. It also took some of the uniqueness out of the special relationship.

The dialogue between the Federal Republic and Israel over weapons began in 1957. During that year the Israeli government approached the Germans for two purposes: firstly to sell Israeli arms to West Germany and later to obtain weapons from the Germnas. It started when in June the Head of the Israel Mission in Cologne, Felix Shinnar, offered the 'Uzi' machine pistol to Franz-Josef Strauss, then the Federal Defence Minister. The Uzi, named after its inventor, an Israeli army officer, was manufactured in Israel and used by the Israeli army. It had proved its usefulness during the Suez campaign. Shinnar was able to tell the Minister that the Uzi was being purchased in respectable quantities by the Dutch army and that other European countries were interested and were examining it.[1]

That the West German Bundeswehr was at that time in the market for purchasing weapons is not surprising when it is considered that until two years previously, before, that is, the coming into force of the Paris Agreements, the Germans had been forbidden by the occupying powers to possess any armed forces. From 1955 therefore the Bundeswehr had to be built up from nothing as rapidly as possible when the West German armament industry had yet to gear itself to satisfy the demands now placed upon it as a result of the country's new international status. Armaments had at first to be purchased as quickly as possible wherever they could be found.

The negotiations which followed had to be kept secret because the Germans feared that there might be strong adverse reactions from the Arabs, but above all because the Israeli government was concerned about the way its public might react to a possible deal resulting in the sale of weapons to the Germans. There was reason to fear that the right-wing politicians in Israel would be strongly opposed for emotional reasons. They still resented any relations with Germany and would have found the sale of arms to that country particularly obnoxious. Some left-wing parties, whose members sat in Ben-Gurion's cabinet, were opposed for different reasons, believing that Israel should maintain a degree of neutrality and disagreeing with Ben-Gurion's increasingly close relations with the West.

The concluded deal was represented on all sides as a purely commercial transaction. The West German Defence Minister later justified the purchase of the Uzi by pointing out that careful consideration had been given to many other similar arms, both German and foreign and that the Uzi was considered to be the most suitable weapon for the Bundesdwehr, offering at the same time the greatest value for money. Shinnar, however, made it clear to Franz-Josef Strauss that export of the machine pistol would not only use up excess production capacity but would also help

Israel's balance of payments.[2] Israel, whose defence expenditure per capita was the highest in the world, desperately needed arms and lacked the funds to purchase them. The export of home-made weapons provided some of the foreign currency to buy those it could not manufacture, as well as facilitate the expansion of its own armaments industry. It must be assumed that on the German side a desire to help Israel financially was also a consideration. Strauss, in his initial talk with Shinnar, had emphasised that he regarded Israel as a stabilising factor in the Middle East.[3] The negotiations over the Uzi pistol showed that the Federal Government, besides continuing to feel a sense of responsibility towards the Jewish State, also had a political interest in its survival and was prepared to aid it in areas beyond those covered by the Restitution Agreement.

Whereas the secrecy of the West German arms deliveries to Israel, negotiated later that year, was preserved at least as far as the general public were concerned, until the mid-1960s – it will be discussed later in this chapter – the sale of Israeli arms became public knowledge in June 1959. A storm blew up in Israel when a report appeared in the West German news magazine *Der Spiegel* that an agreement involving the sale of mortar bombs made in Israel had been signed for the Bundeswehr.[4] Severe criticism was voiced on both the political right and the far left. A cabinet crisis resulted in Jerusalem when some left wing ministers of Ben-Gurion's government, put under pressure by their supporters inside and outside the Knesset, reneged on a cabinet decision which had approved in general terms the decision to give the go-ahead to the sale of weapons abroad. In the end the Ben-Gurion government resigned, but re-formed later without the dissident ministers.

The crisis once again provided evidence of Ben-Gurion's determination not to be defeated by emotionalism when a rational assessment of the situation demanded a pragmatic approach. To the attacks that were made on him by the right-wing opposition and his dissident ministers he had ready answers. He saw no contradiction between the feelings of the Jewish nation and the military needs of Israel. The sales provided dollars and a chance to purchase essential arms for Israel.[5] His reply to opponents who denounced the supply of arms to Germans as a 'defamation of the memory of the six million Jews slaughtered by the Nazis' was that 'the supreme moral of the Holocaust was the need to strengthen and defend the State of Israel'.[6] This transaction, he emphasised, was vital for Israel's security. He argued that Germany had changed since Hitler's downfall to a peaceful middle-sized power, yet had become a member of NATO in which it occupied a prominent position. 'The small State of Israel' he continued,

does not belong to any alliance or bloc. I say this with regret . . . .
Adhesion to a bloc strengthens security, facilitates the supply of
arms and reduces defence expenditure. But we are isolated; we
have to bear by ourselves a great and growing burden of defence
and more than any other nation we need friends . . . .

In an oblique reference to Israel's request for German weapons, Ben-
Gurion added

I do not wish and have no right to discuss here the special equipment
that is needed by the Israel Defence Forces and on which, in my
opinion, our security and survival depend, but I know the places
where we have substantial prospects of obtaining it . . . .[7]

Important supplies of West German war materials are believed to have
started to reach Israel at that time. In an interview with an Israeli newspaper Ben-Gurion said significantly: 'There is a very important reason for
our arms deal with Germany but I cannot reveal it'.[8]

This crisis occurred in 1959, two years after Israeli arms sales to West
Germany had first been negotiated. But all this is anticipating the other
agreement concerning free military deliveries by West Germany to Israel,
which were now in full swing and to which the Israeli Prime Minister,
though obliged to keep them secret, had made some guarded references.
It is necessary, therefore, to return to the origins, two years previously, of
this other agreement.

That the West Germans were in business not only as buyers but also for
selling arms to other countries so soon after they had started to rearm
themselves is not as paradoxical as it may appear. Shinnar writes that
because of the consumer boom and industrial expansion in the late 1950s,
West German industry was not anxious to provide the arms needed and
the Federal Government was obliged to import most of its needs from its
NATO allies. The USA and Great Britain especially were anxious to
export arms to Germany in order to offset the cost of keeping their armies
there, since these countries, unlike the Federal Republic, did not have a
favourable balance of payments. So they sold weapons that were often
old, obsolete, or not entirely suitable, but which the Federal Government
accepted temporarily because of the need to build up the West German
forces rapidly.[9] These were then disposed of as soon as more up-to-date or
suitable material could be obtained. This gave the Federal Republic some
inducement to find a market for obsolete weapons. Many, mainly Third
World, countries were supplied with these and Israel became a potential
beneficiary.

Negotiations between Shimon Peres, then Israel's Deputy Defence

Minister, and Franz-Josef Strauss first took place in December 1957 in Bonn. Complete secrecy was essential to both sides. To Israel because there was opposition in Ben-Gurion's cabinet and in the Knesset to the acquisition, this time, of German weapons by Israel, and it was only after Ben-Gurion had made an impassioned plea in parliament that because of the serious security situation of Israel the government needed to obtain arms wherever it could – and few countries were willing to sell arms to Israel – that the cabinet and a majority of the Knesset, including the left-wing dissidents in the government, agreed that arms could be bought in the Federal Republic if necessary. To the Germans secrecy was necessary, firstly because of the fear that some Arab governments might recognise the German Democratic Republic if a deal were concluded and made public. But there would also have been opposition within the Federal Republic to any arms supplies to Israel, in official circles among those who were concerned about trade with Arab countries, among the general public because it was felt that the Federal Republic should not become involved in Middle Eastern squabbles but should try to entertain good relations with both sides. In the event the details of the Peres–Strauss meeting and the arms deal could not have travelled any further at first than the Foreign Office and from there to the Chancellor. That is why the government Press Office in Bonn, when challenged, was able to report in all honesty that nothing was known there of any Israeli arms purchasing mission visiting the country and that the Federal Government had no intention of supplying arms to areas of tension.

Peres returned to Israel at the end of December without saying where he had been or what, if anything, he had accomplished. It was only ten years later that his partner in the conversation, Franz-Josef Strauss, in an interview with a journalist, revealed some of what had been discussed. He said that the talks had covered not only military questions but also the general relationship between the Federal Republic and Israel. Strauss had declared himself completely in favour of West Germany's policy of Restitution to Israel and its efforts to establish close collaboration with the Jewish State, both of which he regarded as an essential contribution to the Germans' effort at overcoming the past. But Israel was able, as it happened, to offer something in return: not only its experience in the successful military campaign in Egypt, but also information about the large quantities of Russian weapons captured by the Israelis from the Egyptians during the Suez War. There was great interest in these, particularly in NATO, and because of the coolness of American–Israeli relations, the Federal Republic was a useful go-between. Strauss reports on his interest and that of his ministry in the co-operation between Israeli tanks and the Israeli air force during the Suez campaign as well as in captured Soviet war material.

It appears that he obtained the information he wanted. Beyond that he revealed nothing about any practical aspects of the talks.[10]

Peres, interviewed by the same journalist shortly afterwards, said a little more. But from his utterances it would appear that he introduced the general topic of West German–Israeli relations which would have touched upon the Holocaust and Germany's moral debt to the Jews, with a purpose: that of persuading his partners that the Federal Republic ought to contribute to Israel's security, thereby assuring its survival. He maintained that both 'the delivery of German arms to us and the sale of our weapons to the German fighting forces' were discussed. The wording suggests that he was thinking not of Israeli arms purchases, but that the Federal Government should give them free. It seems that no formal written agreement was made; it could not have been made because Strauss was in no position to deliver it there and then without consulting the Chancellor and some of his cabinet colleagues. On the other hand a written agreement would have had to go through the whole government machinery, especially the Treasury, if arms were to be given away free, as well as to the Federal Parliament. Such a formal agreement could not have remained secret.

German military deliveries began in 1959. They consisted at first of non-combatant material such as motor vehicles, training aircraft and helicopters, material needed by Israel for the increase in the mobility of the army. Shooting weapons, anti-tank rockets for example, followed later. Most of this material had to be sent on circuitous routes through sympathetic countries, in some cases as material on loan, to avoid detection. By the end of 1961 the value of West German military equipment sent to Israel had reached DM20 million, a small amount considering what followed later.[11] This period constituted the 'first phase' of the military deliveries; in 1962, following further meetings between Peres and Strauss and a meeting between Adenauer and Ben-Gurion, a new agreement was concluded, leading to a wider range of supplies including some offensive weapons. What is clear from this is that the historic meeting between Adenauer and Ben-Gurion in New York early in 1960 did not initiate military aid given by the West Germans to Israel, as had been widely believed at one time, but rather confirmed it and may have paved the way to an extension of it. What is also interesting is that Peres secretly met Adenauer before the latter's meeting with the Israeli premier.[12] Clearly the new military relationship between the two countries had reached an advanced stage before that meeting.

Because a visit by Adenauer to Israel or by Ben-Gurion to Bonn would have been unacceptable to the Israeli public, the meeting between the Chancellor and the Israeli Premier had to take place on 'neutral' ground.

Even that, it was thought, would be too provocative, unless it could be represented as a meeting that happened by chance. A coincidence had to be found or specially created. The right moment came in March 1960 when both Adenauer and Ben-Gurion were in Washington for talks with President Eisenhower and both moved at one point to New York where they stayed at the Waldorf Astoria Hotel. No advance publicity had, of course, been given by either side, but when it was discovered that the two countries' national flags were flying side by side, the curiosity of the press was aroused. Such a meeting was considered nothing short of sensational.

It was undoubtedly personal motives as well as political considerations which on both sides had favoured such a meeting. Adenauer and Ben-Gurion, despite their different backgrounds, temperaments and positions in the international arena, had certain features in common. In different ways religion played an important role in both their lives; both were the founding fathers of their respective states; both had to start with compromises – Ben-Gurion with a partitioned Palestine, Adenauer with a divided Germany.[13] Both were fighting major rivals in the international field: one President Nasser of Egypt, the other Walter Ulbricht, head of government of the German Democratic Republic. Finally, both were now elderly, having had heroic if largely unsung political careers even before they became heads of their respective governments. The moral courage and tenacity they showed in building a bridge towards each other over a very wide gulf is impressive. It was Ben-Gurion who led a reluctant Israeli people out of the Holocaust syndrome into a more realistic political climate, turning attention away from the past and towards the exigencies of the moment. Adenauer had the courage to demand of the Germans at a time when they were just recovering from a national catastropohe, that they should make considerable sacrifices as a national gesture to atone for the past and redeem the nation's good name. The two eye-witnesses of the meeting say that they were visibly impressed by each other, and both statesmen confirm this by their final statements after the meeting. Ben-Gurion had repeatedly said that 'the Germany of today' under Adenauer's leadership 'is not the same as the Germany of yesterday'. He found himself confirmed in his view. Although he received nothing but verbal assurances from Adenauer, he clearly trusted him enough to be satisfied that they would be followed up.

What is known about this meeting is that in general terms Adenauer agreed that the Federal Republic should continue to aid Israel with free arms supplies but no details were discussed and no formal agreement was made at the time. The details were left to Peres and Strauss. All the Israeli Prime Minister needed was Adenauer's assurance that he was behind his Defence Minister. This he obtained. On the question of economic aid which he also raised at the meeting, Ben-Gurion mentions some aspects in later interviews. He had proposed in New York that the Federal Republic might grant Israel development aid additional to the restitution provided under the Luxembourg Agreement which still had some years to run. His idea was that the bulk of this aid might be used to develop the Negev Desert which formed almost half the land area of Israel, an undertaking that was dear to the heart of the Israeli Premier and which he hoped would be attractive also to the Chancellor and one day to the German public. Adenauer gave Ben-Gurion to understand verbally that he agreed with this proposal also. Here, too, no documents were signed and no details were discussed. The communiqué that followed the meeting only made brief reference to 'aid for Israel'. As in the case of the military supplies, Adenauer had simply given his word, but like the agreement on arms, it was to be kept secret.

As in the case of the Luxembourg Treaty, several reasons have been given for Adenauer's continued generosity towards and interest in the State of Israel. They generally relate the arms and economic aid agreements to the national interest or current pressures in day-to-day politics. It has, for example, been suggested that the swastika daubing which had occurred in many German cities only two months before the Adenauer–Ben-Gurion meeting could have pricked Adenauer's conscience. Others suggest that the unwillingness to establish diplomatic relations with Israel could have had this effect. It is improbable that the Israeli Premier, though he was anxious to normalise relations with Bonn, raised this matter with the Chancellor who would certainly have refused. But these are insufficient reasons for doing a deal of the dimensions indicated and the risks involved. Nor are there any signs that the Americans pressurised Adenauer at that stage into either supplying arms or making more economic aid available to Israel, though it is likely that the Chancellor would have consulted them about the possibility of West German arms or NATO equipment from the Federal Republic being secretly sent out there and that the United States government, concerned about the build-up of Soviet weaponry in some Arab countries, would have supported him. There seems little doubt, on the other hand, that moral considerations again played a large part in Adenauer's action and that he considered that the Germans owed a moral debt to the Jewish State which required the Federal Republic, notwithstanding some political risks, to continue aiding it for some time to come.

The second stage of West German arms deliveries, already referred to in this chapter, began in 1962 following the Adenauer–Ben-Gurion meeting and intensive negotiations between Strauss, Adenauer and Peres. Deliveries still consisted mainly of surplus Bundeswehr material, although

American helicopters have been mentioned.[14] The final and most important stage began in 1964 and this time at the request of the United States government. When Khrushchev, in May of that year, had made promises of more weapons to Cairo, the Israelis appealed to Washington for more military aid. The United States government, worried about Arab reactions if this request were granted, directed them to the Federal Republic. Erhard, who had succeeded Adenauer as Chancellor the previous year, visited Washington in June, and was pressed to send US M-48 tanks to Israel. West German qualms were apparently soothed by promises of US moral and financial support if this led to Arab counter-measures.[15] A much wider range of arms and material was now due to be sent from the Federal Republic to Israel; they are said to have included, apart from those already mentioned, 50 aircraft including Noratlas transporters, Dornier DO-27 communication planes and French 'Fouga Magister' jet trainers, anti-aircraft guns with electronic homing devices, helicopters, howitzers, submarines and speedboats.[16] All came from NATO stock held by various countries – the two submarines came from the United Kingdom, the AA guns from Sweden, the helicopters from France – but all were sent in the name of the Bonn government. Some were, for the sake of secrecy, trans-shipped via third countries. In the case of the tanks the empty shells were sent to Italy, followed by the 'inerts' which were then fitted so that the complete tanks could be shipped from Italy to Israel.[17] The total value has been estimated at DM240 million, but because the weapons were all 'used', this figure may be very wide of the mark. It is not known, moreover, how much of the promised material remained undelivered when the consignments abruptly came to an end in the spring of 1965.

Israel was not then the only country to receive arms from the Federal Republic. Weapons were sent from there to other NATO countries as a matter of course. But the Federal Government also sent them overseas to non-aligned developing countries. Arms deliveries to these states were, like economic aid, an inducement not to establish relations with East Berlin, as these deliveries could be withdrawn in case of non-compliance. But because West German arms were delivered to several countries and because these deliveries were for the most part secret, a machinery existed for by-passing the usual governmental and administrative agencies in the Federal Republic. Funds were thus available for arms deliveries, some of them coming from sources other than the defence budget. It was therefore not necessary to pass legislation. As will be seen later, the complete lack of parliamentary control over this type of expenditure was severely criticised by politicians and by the media once the secret of arms deliveries was revealed and a crisis in West German foreign relations had ensued. Six members of the Bundestag, who were initiated, were said to have had mixed feelings from the beginning: the two members of the Social Democratic opposition were particularly critical, for both ideological and practical reasons, about the Federal Government sending arms to the Middle East. They would have advocated the strict adherence by their government to what was, in fact, official West German policy: that no weapons should be sent to areas of tension. But because the agreement had already been concluded without consultation, and because total secrecy had been a condition of their being informed, the dissenting parliamentarians were unable to act.

The agreements on arms and economic aid for Israel highlight the contradictory nature of West German foreign policy and the impasse into which the Bonn government's Middle Eastern policy was being manoeuvred after 1957. The risk taken when a country courts the friendship of two parties implacably hostile towards – indeed technically at war with – each other is obvious and there were warning signals as early as 1952 when the Luxembourg Restitution Agreement was signed. Once this is extended to military aid being given to one side, which is known to need it to fight the other, then serious complications can be expected unless by a miracle such aid can be kept totally secret for a long period. This certainly was not the case here. That the Federal Republic was itself engaged in a bitter conflict, resulting from the division of Germany and Bonn's struggle to keep the other German state ostracised in the international community, did not help matters. It merely opened another flank for the Federal Republic, which Israel's enemies, the Arab states whom the West Germans had regarded as their friends, could attack.

To put it more drastically, it at worst exposed the Federal Government to blackmail or, at best, threatened to subject its Middle Eastern policies to unacceptable Arab pressures. The period of the arms deliveries can be seen as the height of the conflict between West Germany's moral debt and its other interests in the international political system. Thereafter this conflict does not simply go away but, to borrow a phrase, moderates and changes direction as West German foreign policy begins to change. On the other hand the arms deliveries led to a marked improvement in relations between West Germany and Israel at governmental level – but not including the Israeli public for they were unaware of what was happening – to the extent that despite the Israelis' preoccupation with the past, the Israeli government exercised unusual patience over other disagreements with Bonn at that time. These will be the subject of the next chapter. The storm, when it broke late in 1964, had to be weathered by Adenauer's successor, Ludwig Erhard, together with a surprised and bewildered German public who were none too happy with the aftermath. This will be described in a later chapter.

NOTES

1. Shinnar, op. cit., p. 139.
2. Shinnar, op. cit., p. 139.
3. Shinnar, op. cit., p. 140.
4. *Der Spiegel*, 25 June 1959.
5. *Neue Züricher Zeitung*, 28 June 1959.
6. *Israel Digest*, 10 June 1959.
7. *Israel Digest*, 10 July 1959.
8. Interview with *Yediot Acharonot* quoted by *Le Monde*, 5/6 July 1959.
9. Cf. Shinnar, p. 141.
10. R. Vogel, op. cit., p. 137.
11. *Deutschkron*, p. 29.
12. R. Vogel, p. 142.
13. Cf. *Deutschkron*, p. 126.
14. *FAZ*, 20 Feb. 1965.
15. Ibid.
16. Ibid.
17. Ibid.

5

German Scientists help the Egyptians to Prepare for War

One of the factors that profoundly disturbed West German–Israeli relations in the early 1960s was the employment of German experts in the armaments industry of Nasser's Egypt. Their presence in the Middle East was partly a hang-over from the Second World War: a number of confirmed Nazis who had occupied prominent positions in the Third Reich fled after the collapse of Hitler's empire to Latin America or the Middle East, most of them to escape from Allied or West German justice, some in the hope no doubt of preserving the ideals of Hitlerism. In Cairo they formed the nucleus of a German colony that grew when Egypt increased its economic links with the Federal Republic and when the Egyptian government offered jobs to foreign scientists who could help the country develop its own sophisticated weapon systems. This policy was in line with the Egyptian government's stance of non-alignment, the aim of which was to preserve the greatest possible independence of the two great power blocs. But Nasser's declared aim was also the destruction of Israel. The German scientists who were working for the Egyptian government in the military field – whether they were unrepentant Nazis or had simply been attracted by work prospects such as they would not find in their own country at that time – must have known they were helping Nasser's Egypt in preparing for a war against the country which had given shelter to hundreds of thousands of survivors of Hitler's policy of extermination. The Israeli government, perceiving a threat from a possibly more advanced Egyptian war machine, persistently protested to the Federal Government about the activities of its citizens in the Middle East. The Federal Government, though deeply embarrassed, maintained that it had no power to prevent its citizens from going to work for foreign governments or to recall those already there. The two questions one must ask when dealing with this episode are whether the work of the German scientists in Egypt really constituted a serious threat to the Israelis and why the West German government appeared to do nothing to stop them.

The effect that home-developed and produced weapons could have had on Nasser's chances of winning technological superiority over the Israelis must be seriously doubted in view of the Egyptian leader's ability to obtain weaponry both from the Soviet Bloc countries and the West. These doubts are increased when the qualifications of the German scientists working for him are examined. Some of them had, it is true, been employed at Hitler's rocket research stations, such as Peenemünde, which developed the flying bombs and rockets fired at British cities towards the end of the Second World War. But none of these scientists was said to have achieved eminence. After the end of the war there were many opportunities both in the West and in the USSR for the development of rocket technology and the top-ranking scientists found employment, especially in the USA. Those who were second-rate did not, neither did they have opportunities at first in their own country, since the Federal Republic was until 1955 forbidden to manufacture weapons or to have any under its control. When the offers came from the Third World, therefore, some scientists seized the opportunity. As the West German constitution guaranteed freedom of movement to all its citizens, there was no reason why these scientists should not move to Egypt or commute between Cairo and West Germany as and when it suited them. Surprisingly, the Egyptian leader seems not to have been short of funds for weapons research despite his economic difficulties.

German scientists are known to have been active in Egyptian weapons research and production since the early 1950s, but it was not until 1958 that substantial projects were under way and a number of German citizens were engaged on them. The Egyptian president was then building up his military power in order to fulfil his international political ambitions: these included not only the destruction of Israel but also the domination of the Arab world and his establishment as the leader in the struggle for Arab liberation. Such a position alone would require military intervention in Arab affairs at some stage. Egypt, though of all the Arab states the largest in population and the most technically advanced, was nevertheless not in a position to challenge Israel's technical superiority in the Middle East. Nasser knew that he could not rely only on the Soviet Union where he was unable to obtain the most up-to-date weapons, and though he was close to Moscow at times, he was unwilling to abandon his country's non-alignment. Hence his efforts to establish an Egyptian arms industry with the help of foreign experts.

The link between the Egyptian government and the German weapons researchers was a Swiss arms dealer of Egyptian origin, Hassan Sayed Kamil, who had worked for the Swiss arms firm Oerlikon. Kamil's tasks were to recruit scientists for the Egyptian government and to buy machines and parts for the weapons to be produced in Egypt. As much of this material had to come from West Germany or Switzerland, both of which had imposed an embargo on arms to the Middle East because of the troubles there, he had to channel its goods by devious international routes. It was Kamil who approached veteran aircraft designer of World War Two fame, Willy Messerschmidt, whose company Hispano Aviacion had developed under his guidance two supersonic fighter planes, one of which was capable of flying at twice the speed of sound. The project was nearing completion when progress was slowed through lack of funds. The injection of Egyptian money, however, soon led to the completion of the two aircraft, the H 200 and H 300, which are believed eventually to have seen service in the Egyptian air force.[1]

Messerschmidt did not settle in Egypt and denied that his companies had been directly linked with the Egyptian weapons and aircraft research centres, but he travelled to Egypt frequently and gave advice. A similar role was played by a German rocket expert, Dr Eugen Sänger, who headed the Institut für die Physik der Strahlantriebe in Stuttgart and during the Second World War had worked on flying bombs and rockets at Peenemünde. By 1960 he too was commuting to Cairo. When the Egyptians were offering part-time employment in the manufacture of 'space rockets for meteorological purposes' some of Sänger's collaborators at Stuttgart, frustrated by the limits imposed by the Federal Government on rocket development in the Federal Republic, accepted.[2] Of this the Israeli secret service received news. The Israeli government protested, whereupon the Federal Government demanded the dismissal of the Stuttgart 'part-timers'. Several of the others, however, now established themselves more permanently in Cairo, though in some cases continuing their shuttle between the Federal Republic and Egypt.

Two years after the events in Stuttgart an international sensation was caused when in July 1962 two new types of rocket were publicly paraded through the streets of Cairo. German scientists, it became known, had been responsible for the development of these rockets, one of which was thought to be capable of travelling 360 km and could easily have reached Israeli cities.[3] The Israeli reaction was not just one of dismay that their technical superiority was being challenged by the Egyptians – and to Israel, a much smaller country than Egypt both in area and population and therefore dependent for survival on this superiority, this was a serious matter – but also one of horror that German citizens were again at the forefront of producing the means that could destroy a large section of the Jewish people, thereby continuing the work of genocide begun by the Third Reich. The combination of the threat to the country's security and the memories of the past reached an emotional level higher even than that created by the anti-Semitic upsurge in 1959–60 and the

Eichmann trial in 1961. The subject of the German scientists working for the Egyptian government became a major issue in Israel over a period of three years. It was frequently featured in the press, became the subject of many Knesset debates and government statements and led to government action which took various forms, from official protests to the Federal Government to assassinations by undercover Israeli secret service agents. It caused a sharp deterioration in West German–Israeli relations, despite the continuing flow from Bonn of restitution and the secret arms deliveries.

In the short term the Israeli government could only draw the matter to the attention of the Federal Government – which it knew not to be responsible – and ask it to intervene. The West German government said in reply that it would 'do everything possible to persuade the scientists to return to Germany',[4] but gave no hint of any official action that might compel the experts to cease their activities. The Israeli government therefore decided to take direct action.

The matter began to attract worldwide attention early in 1963 when a German rocket scientist, Dr Hans Kleinwächter, was shot at as he was driving home from work at Lörrach, a small town near the Swiss–German frontier and close to Basle. He escaped injury. He had frequently commuted between his home town and Cairo, to which his company is reported to have exported electronic equipment for the Egyptian rocket industry. The assassination attempt was linked with the unsolved kidnapping of Dr Heinz Krug in Munich some weeks earlier.[5] Apart from an anonymous letter that he was dead, the mystery of Krug's disappearance was never solved.[6] He, like Kleinwächter, had been a rocket scientist working for the Egyptian government. Both scientists were reported to have worked on the two rockets demonstrated by the Egyptians the previous year under Professor Sänger.[7] He left Egypt eventually under strong pressure from the Federal Government, when the German authorities became embarrassed about the adverse international publicity given to their scientists in Egypt.

Other incidents that had occurred and were related to these events soon came to light. In July 1962 an aircraft chartered by Meco, the Swiss firm owned by Hassan Sayed Kamil, had crashed in Germany. Kamil, because of a last minute switch, was not on the plane, but his wife was killed. The crash was never explained.[8] In November of that year a number of letter and parcel bombs bearing Hamburg postmarks were sent to Heluan, the site of the Egyptian rocket station. Five Egyptian workers were killed and a German secretary to one of the scientists was seriously injured.[9] When finally, in March 1963, two men were arrested in Basle for threatening the daughter of another scientist who was then working in Cairo, it became clear that the various incidents had been the work of the Israeli secret service.

The two Israeli agents arrested, one an Austrian, the other from Tel Aviv, had phoned Miss Heidi Goercke, daughter of Professor Paul Goercke, an expert on electronic steering, at her home in Freiburg, southern Germany, and asked her to meet them in Zurich if her father's life was worth anything to her.[10] She kept the appointment and was urged to go to Cairo, where her father was working on rockets for the Egyptian government, and persuade him to return home. The request, so Miss Goercke alleged, was backed by threats to her and her father's safety. The girl, who was aware of the dangers then threatening the rocket experts, informed the police and helped to lead the two Israeli agents into a trap.

During the investigations that followed it became clear that the two men had been involved in the kidnapping of Krug and the attempt on Kleinwächter's life. One, a Dr Joklik, was said to have been an expert on radioactive isotopes and to have worked at the Egyptian Institute of Advanced Technology in Cairo.[11] He said at his trial that he had stopped working for Egypt when he realised that strontium 60 and radioactive cobalt were being obtained by Dr Wolfgang Pilz, the leader of the German scientists in Egypt, for loading into rockets capable of destroying Israel.[12] The implication was that he had left the service of the Egyptian government for moral reasons and decided to give his services to Israel instead. At the time Miss Goercke failed to repeat before the court a remark she had made previously that Joklik had threatened her father's life. Joklik and his fellow agent, Joseph Ben-Gal, an Israeli secret service man, each received a three-month prison sentence counted against their time in custody, and were freed. The lenient sentence was justified according to the Swiss public prosecutor by the seriousness of the evil of the weaponry produced in Egypt and the threats by its leader to use them.[13]

The activities of the Israeli secret service demonstrated the seriousness with which the Israeli authorities regarded the aid in modern weapons technology given to the Egyptian government by the West German scientists. They were stopped, however, after the arrest of Joklik and Ben-Gal on the orders of the Israeli Prime Minister when it became clear that they had begun to cause anger and resentment in West German official circles and among German friends of Israel, and that Ben-Gurion's carefully nurtured relations with the Federal Republic could be adversely affected. The Israeli government thereupon decided to stop undercover action and to bring the matter to world attention instead.

In a rare display of national unity the Knesset passed without a dissentient voice a resolution asserting that the activities of West German scientists and experts working in Egypt on the manufacture of weapons of mass destruction were a grave danger to the security of Israel and its population, that the West German government must put an immediate end to these dangerous activities of its citizens and that the enlightened world should exercise its influence to stop them. The resolution followed a statement by Golda Meir, then Foreign Minister of Israel, in which she said that to the efforts of the Egyptian leader to build up his country's strength in order to destroy the State of Israel a new element had been added recently: a number of German scientists and hundreds of German technicians were helping to develop offensive missiles in Egypt and even armaments banned by international law which served solely for the destruction of living beings. The Egyptian government was endeavouring to obtain these types of weapons which other powers were not prepared to supply to Egypt

> through groups of conscienceless scientists, who are not only prepared to do Egypt's will but freely make their own contribution to the same aim. There is no doubt that the motives of this evil crew are, on the one hand, lust for gain and on the other a Nazi inclination to hatred of Israel and the destruction of the Jews.

Israelis could not accept the Federal Government's explanation that legal difficulties stood in the way of preventing the scientists from acting as they did and demanded that the West German government put a stop to their activities, if necessary by legislation. She believed that Israel was supported in this demand by world public opinion.[14]

The allegations that German citizens were engaged in the manufacture of weapons of mass destruction in Egypt were strongly denied by the Bonn government. Two days after the Knesset resolution a Federal Government spokesman claimed that according to information received by Bonn a maximum of 11 German scientists were engaged in missile development in the United Arab Republic, and that a large number of West German, Austrian, Swiss and Spanish nationals were working at a Swiss–Spanish aircraft factory there. The Federal Government, the spokesman continued, disapproved of West Germans engaging in armament production in areas of conflict, but had no means of preventing them from leaving the country unless they had violated the laws of the Federal Republic or of the host country. Nor could the government take measures which would restrict the freedom of movement guaranteed under the Federal Constitution. In a further statement some days later the spokesman added that enquiries by the West German embassy in Cairo had shown that there was no evidence of Germans being engaged in Egypt on the production of atomic, biological or chemical weapons. The Federal

Government would, however, endeavour to bring about the return of West German citizens whose activities abroad might contribute to an increase of political tension and was investigating whether such activities could be effectively prevented by further legislation or administrative measures.[15]

While West Germans were undoubtedly involved in the development of Egyptian iron and steel, rocket and aircraft industries, the more sensational reports about their part in the production of weapons of mass destruction must be discounted. With or without the existence of a nuclear reactor, the Egyptians would not have had the funds or the facilities to develop nuclear warheads at that time. The military value of some of the other weapons, including the cobalt warhead – even though it was claimed by Joklik at his trial that the Egyptians were importing cobalt in quantities in excess of what they needed for medical purposes – must be doubted. If Nasser intended to produce aggressive weapons, he must have had other priorities. In any case, no evidence was found of any use of the type of weapon described above in Nasser's later military campaigns. If, on the other hand, some of these stories were not worthy of credence, then neither were the statements by Professor Pilz, one of the chief scientists, that his men were engaged only on rockets intended for meteorological purposes. The question was asked pertinently whether these were needed sufficiently to warrant the expenditure of the none too plentiful Egyptian resources.[16]

In the uncertainty the Israelis undoubtedly overreacted. The fear that highly sophisticated weaponry could fall into the hands of the leader of a state who had repeatedly vowed that he was working for the destruction of the Jewish State was natural. But what heightened Israeli emotions and resentment was that some of the participants were Germans. By the early 1960s much of the deep hatred felt towards the Germans in Israel had abated. As a result relations between the two peoples had improved at all levels. Even the Eichmann trial in 1961, with its many revelations of horror and crimes committed during the German occupation of eastern Europe, had not been able to awaken such strong emotions. Now old antipathies began to manifest themselves again. 'The Germans must recognise that Israel cannot stand idly by while Germans build rockets for Nasser which are intended to destroy Israel', wrote the *Jerusalem Post*.[17] And *Ha'aretz* said that the Federal Republic was to blame if Israel should be forced to take 'unconventional measures in order to defend itself against the threat of unconventional and cruel weapons', while *Ma'ariv* accused the Federal Government of not adequately condemning the activities of the German scientists in Cairo. Newspaper readers were reminded that Germans were again active in attempts to destroy the

Jewish people.[18] Above all there was incomprehension that the West German government should continue to drag its feet and not take the necessary action to bring its nationals to book or, at the very least, to force them to come home.

These were moderate statements. There was, however, considerable preoccupation in the Israeli press with reports about weapons of mass destruction being produced by Egypt with the help of German scientists. The reports seem to have been encouraged by the Israeli secret service with the object of creating a worldwide furore, thereby reinforcing and at the same time justifying the campaign of physical violence or threats of violence against the German scientists. It was probably hoped to frighten them into giving up their work in Egypt and in some cases it may have been effective. It was at that stage that Ben-Gurion intervened with the Chief of Israeli Security, Isser Harel, fearing adverse developments in Israeli–West German relations. On 25 March 1963 the Security Chief resigned after being told by the Prime Minister that he, Ben-Gurion, could not 'accept his evaluation'. This is believed to have referred in particular to reports spread by the security services that the Federal Government had tolerated the development of chemical and bacteriological weapons by West German scientists in Egypt because such weapons were banned in the Federal Republic.[19] Ben-Gurion once again showed that he had a greater overall view of his country's situation and a more realistic assessment of the danger that threatened Israel through the activities of the West German scientists. He did not underestimate the problem. That is why at the height of the anti-German press campaign in March 1963 he invited the Secretary of State at the West German Defence Ministry, Volkmar Hopf, who was in Israel at the time, for a talk at his holiday residence in Tiberias. It is believed that he made no secret of his concern about the activities of the West German scientists in Egypt and that he and Golda Meir asked Hopf to convey their views to the Federal Government.[20] But he knew that his government could not afford to risk a conflict with the Federal Republic as long as Israel was receiving large quantities of arms and war material through the good offices of the West Germans. He and the military establishment in Israel would have assessed the value of limited sophisticated weaponry in the hands of the Egyptians, who lacked the knowledge to use them effectively, as not high. It had been proved during the Suez campaign that the Egyptians were not very competent in the use of their weapons, with which they had been supplied by the Soviet Bloc. The Israelis' superiority had lain precisely in their skill at using highly technical equipment and the ability of their military leaders. This was to be proved again in subsequent Arab–Israeli conflicts. In that situation therefore, the supply of much needed arms from West

Germany must have had priority over whatever contribution West German scientists were making to Egyptian armaments. If Ben-Gurion had a problem, it would be that he could not explain this to his public without revealing an important secret.

Ben-Gurion's attitude to the problem of the West German scientists was therefore widely disapproved of in Israel, even by some members of his cabinet, and caused a political crisis. But at a meeting called by the opposition the Knesset rejected by 67 votes to 47 opposition proposals to debate the activities of the West German scientists in Egypt and the resignation of the Security Chief.[21] Without belittling the activities of the scientists Ben-Gurion was openly critical of the anti-German press campaign in Israel. He accepted an opposition charge that he had suppressed some documents about the extent of West German assistance in the arming of the United Arab Republic and added that he had caused the press campaign to be toned down. He said

> To my great regret a large number of falsehoods and distortions have been published in recent weeks in the local press, some of them, I assume, in error, but in several cases on purpose and without a sense of responsibility .... Our grave concern over the designs of the Egyptian dictator to destroy Israel and the assistance he is receiving from German and other scientists and technicians should not throw us off our balance. For obvious reasons it is not desirable that the debate should take place in parliament for there are certain things that cannot be spoken of here.

He also denied that he had dismissed the Head of Security Services.[22]

Were the West Germans really doing nothing to stop their compatriots boosting Nasser's arsenals in Egypt? Not if we are to believe Felix Shinnar: 'Through direct, unofficial approaches, by way of quiet diplomacy the German side did more than the public knew about.'[23] He does not elaborate. On the other hand there were at first few outward signs that the West German authorities were acting. The Federal Government was indeed in a painful dilemma: the Basic Law, like similar parliamentary constitutions elsewhere, guaranteed freedom of movement to all West German citizens. The government could not touch its citizens unless they had broken West German law. They had not done so. The West German government, backed by politicians, the media and intellectuals, having since the war extolled the virtues of the *Rechtsstaat*, the state governed by the rule of law, and having been at great pains to assure the West German public and the world at large that the Federal Republic would respect democratic freedoms and the rights of the citizen, could not now break its own constitution in order to provide for a solution to one specific problem, however embarrassing this may be. There now began therefore a frantic search by West German constitutional lawyers, Bundestag members and other politicians for a way out. One answer seemed to be provided by Article 26, paragraph 1 of the Basic Law which says that 'activities tending to disturb . . . peaceful relations between nations, . . . or preparing for aggressive war are unconstitutional'. Surprisingly arguments based on this failed to convince West German constitutional lawyers, as did other arguments in favour of new legislation.

In September 1962, six months before the Knesset resolution, Professor Böhm, a CDU member of the Bundestag, one of Adenauer's negotiators during the period leading up to the Luxembourg Agreement and an ardent supporter of Israel, was received by Golda Meir in Jerusalem on the occasion of the tenth anniversary of the signing of the agreement. The Israeli Foreign Minister informed Böhm of Israel's concern over the activities of the West German scientists in Egypt and asked him to use his influence to have them stopped. In May 1963 Böhm, at the head of the Bundestag committee comprising all three parties in the West German parliament, submitted a bill which would have made the work of West Germans on atomic, biological and chemical weapons subject to permission by the West German Foreign Ministry and failure to do so a punishable offence. Those already working on such projects were to obtain a retroactive permit within six months or be arrested on returning to the Federal Republic. The proposal was dropped because of objections within the CDU that it would be ineffective, while others claimed that it contravened the right of freedom of movement. The Bundestag members of all three parties therefore unanimously passed a resolution requiring the government to submit alternative legislation. This was not done, nor was the matter followed up by the Bundestag members. A later attempt by an inter-ministerial committee representing the Foreign, Interior, Economic and Justice Ministers to introduce a passport law which would have led to the withdrawal of passports from scientists who continued their work without a government permit met with the same constitutional and legal objections and had to be abandoned. The – mainly moral – arguments that the German scientists were helping to prepare for an aggressive war, or that a possible victim would be Israel, to which the West Germans owed a special moral debt, that some of the scientists were Nazis and wished to continue the fight against the Jewish people, or the simple point that their activities were damaging to the international reputation and therefore the interests of the Federal Republic were of no avail against the overriding claim by some members of the Bundestag and the government that any attempts to introduce legislation restricting the activities of the scientists would clash with a basic principle embodied in the constitution. Even Article 26 of the Basic Law, which made the creation of dissension among nations or the contribution to preparations for an aggressive war a punishable offence could not override the principle. Attempts to amend the Article or to extend it also failed.

The Federal Government finally took the line that in the absence of legislation it would try to exert a moral influence on the nation. In an interview with *Deutsche Welle* on 21 April 1964 the Minister of the Interior, Höcherl, put it this way:

> We should like to exert more moral influence on our people, so that they understand the attitude of the public and of parliament. They should not even begin to succumb to the temptation of involving themselves in such activities for business or professional reasons. They should esteem peace, the greatest good, so highly that they are not prepared to exchange it even for the greatest personal advantage.

A laudable statement, perhaps, but unconvincing. When in a later debate two SPD members asked the same minister whether apart from legal reservations he also had political misgivings in view of possible repercussions in Arab countries, Höcherl gave an evasive answer.[24] More surprisingly the Foreign Ministry, asked by two prominent SPD members of the Bundestag about the possibility of passing a law which would deprive West Germans of the ability to produce aggressive weapons in Egypt, also adopted a negative attitude.[25]

The statements of the two ministries confirm the suspicion that legal and constitutional objections were not the only, perhaps not even the main, reasons for the Federal Government's hesitation. Even if it is assumed with hindsight that a compulsory withdrawal of the scientists by government action would not have led to Egyptian recognition of the GDR, this could not at that time be regarded as certain. Another point is that Nasser by 1963 almost certainly knew of the secret West German arms shipments to Israel. If he did, and if Bonn was aware that he did, this would have been added reason for West German nervousness. It is not known why the Egyptian President did not react immediately when he found out, but bided his time for another year. He may have waited for an opportune moment. But such a moment could easily have been created if the West Germans had thwarted Nasser's Egyptian rearmament programme while continuing the secret arms deliveries to Israel. There is finally the question of Bonn's role of helping to keep the USSR as far as possible out of the Middle East. Nasser, anxious to save his rearmament programme, could well have turned in desperation to the Soviet Bloc to replace those West German scientists who would have been forced to leave. There were signs that the American government discouraged the taking of legal or administrative measures by the West Germans for that reason. The *New York Times* reported from Washington the State Department's fear that if the West German scientists were withdrawn they might be replaced by Soviets while West German help with aircraft production reduced Cairo's reliance on the Soviet Bloc. According to Averell Harriman, then American Under-Secretary for Political Affairs, Cairo was, by employing western sources, 'seeking to reduce its almost total reliance on the Soviets for military equipment'.[26] All this suggests that there were good international political reasons for Bonn's reluctance to legislate. The legal arguments, on the other hand, are greatly weakened when it is considered that there was a ready-made Article in the constitution making activities tending to disturb peaceful relations between nations or preparations for aggressive war unconstitutional.

So we come back to Shinnar's reference to 'direct unofficial approaches' and 'quiet diplomacy' by which 'the German side did more than the public knew about'. The West Germans discreetly recruited their scientists back to the homeland by 'exerting moral influence', reinforced by substantial material inducements. Discretion was essential to avoid arousing the anger of the Egyptians. It was a gamble but it eventually paid off because of technical developments in the Federal Republic. What seems in the end to have brought most of the scientists home was, apart from moral pressure, lucrative offers of alternative employment in West Germany for rocket experts as the Federal Republic entered the European satellite business, a genuinely peaceful enterprise.

In Israel, where Ben-Gurion was soon replaced by Levi Eshkol as Prime Minister, anger with the West Germans simmered on for a while. Because of the slow process of bringing the scientists home and the secrecy which surrounded this and other matters affecting the relationship, the Israelis accused the West Germans of procrastination and insensitivity. Once the secret of the West German arms deliveries to Israel was revealed, however, the dispute over the West German scientists was overtaken by other more important events and Israeli ill-feeling subsided. Most of the scientists had in any case returned home by then. This will be dealt with in another chapter.

NOTES

1. *Frankfurter Rundschau*, 'Das Geschäft mit Herrn Nasser', by Peter Miska, 10 Aug. 1963.
2. *Deutschkron*, op. cit., p. 232.
3. *Deutschkron*, op. cit., p. 233.
4. Shinnar, op. cit., p. 134.

GERMAN SCIENTISTS HELP THE EGYPTIANS

5. *New York Herald Tribune* (Paris) 23 & 24 Feb. 1963.
6. *Der Spiegel*, 1963, no. 13, p. 68.
7. *The Observer*, 24 Feb. 1963.
8. *New York Herald Tribune* (Paris) 23 & 24 Feb. 1963.
9. *Der Spiegel*, 1963, no. 13, p. 68.
10. Ibid.
11. *FAZ*, 21 March 1963.
12. *The Guardian*, 10 June 1963.
13. *Daily Telegraph*, 12 June 1963.
14. *Israel Digest*, 29 March 1963.
15. *Keesing's Contemporary Archives*, p.19635A.
16. Cf. *The Observer* report, 24 June 1963.
17. *Jerusalem Post* quoted by *Der Spiegel*, no. 13, 1963, p. 70.
18. Quoted by *Deutschkron*, op. cit., p. 241.
19. *Keesing's Contemporary Archives*, p. 19635A.
20. *Deutschkron*, p. 346.
21. Isser Harel was known to have been instrumental in the kidnapping of Eichmann in Argentina in 1960. His sudden resignation therefore came as a shock to the Israeli public.
22. *Israel Digest*, 29 March 1963.
23. Shinnar, op. cit., p. 134.
24. Jekutiel Deligdisch: *Die Einstellung der Bundesrepublik Deutschland zum Staate Israel* Bonn-Bad Godesberg: Verlag Neue Gesellschaft, 1974, p. 75.
25. Jekutiel Deligdisch: ibid. (personal interview of author with Professors Carlo Schmid and Franz Böhm).
26. *New York Times*, 12 April 1963.

6

The 'Overcoming of the Past' and the Prosecution of War Criminals

The 'overcoming of the past' is a phrase coined by the West Germans themselves to describe the efforts to be made by the German people as a whole to live down, as it were to purge itself, of the heavy moral burden weighing on it as a result of crimes against humanity, committed in its name by agents of the political establishment of the Third Reich. The subject is vast and goes too far beyond the scope of this book to be treated in full here. In addition to the six million Jews, millions of non-Jewish civilians were murdered by the Nazis in occupied Europe, including in Germany itself. But some treatment of this subject is called for as it has always touched and continues to touch a raw nerve in West German–Israeli relations.

The payment of a large sum of compensation to Israel under the Luxembourg Agreement was motivated by the desire of the Adenauer government to make material amends, and yet it was always stressed that material restitution could never completely repay such a debt. It was always clear that the 'overcoming of the past' could in no way be achieved by the payment of money though it is not easy to define what mental and psychological processes would be necessary in order to throw off the burden of the past. While on the whole German post-war policies have been characterised by financial generosity, not only towards Israel and the Jews, the attitude of the German public including that of some politicians and other public figures to what happened during the period of 1933–45 has often come in for severe criticism.

The debate which ensued in Germany after the war, after the extent of the mass killings and tortures by the Nazi regime had been revealed, centred on the question of collective guilt of or complicity in those crimes. Most Germans have not accepted the notion of 'collective guilt' of the whole German people, nor does it seem reasonable that the ordinary citizen who was not involved in the perpetration of crimes and could not have prevented them, can be regarded as guilty, whether he knew they were being perpetrated or not. What is generally recognised as essential, however, is a sense of responsibility, a need for both the individual and the collective to associate with the crimes that were committed, an awareness that they were committed by a German government and on behalf of the German people as a whole, a government which had loudly proclaimed that it was acting in the name of the people and that it was supported by a vast majority. To do otherwise, to dissociate oneself from these events, to disclaim responsibility on the grounds that one was not actively participating either in the killings or in shaping the events that led up to them is a renunciation of moral principles. The denial of moral responsibility and what is even worse, an attempt to play down the gravity of the acts by claiming they were accidents of war or had been committed on the orders of a superior, or could be excused or condoned for any other reason, are dangerous in that they may lead to a repetition of the crimes.[1] Attempts to seek refuge by 'forgetting' or calling upon others 'to forget the whole unpleasant business' are not the answer. While constant recriminations on the one hand and a perpetual morbid self-contempt on the other are undesirable, only a constant awareness and vigilance can prevent such happenings recurring in the future.

These are hard prescriptions for ordinary men and women, most of whom will have been as overwhelmed by the horrors committed by the Nazis as were ordinary decent people all over the world. But critics both inside and outside Germany have noted a failure on the part of some German people and many of its public figures to adopt the right attitude towards the Nazi past. While German governments have been correct and even generous in making material restitution to those who directly or indirectly suffered, there were signs that in some other respects the process of 'overcoming the past' was failing. Three interrelated factors, which have been regarded as a test of the German will to overcome the past, have been the nation's success in eradicating all traces of the Nazi ideology, at least from public life; the elimination of former active Nazis from public posts; and the punishment of those who during the 12 years of the Third Reich were planning or participating in the killing or torturing of innocent people. It is the third of these categories that is to receive special attention in this chapter as one of the important factors that influenced the development of West German–Israeli relations.

The problem of bringing the Nazi war criminals of the Second World War to justice is complicated by many factors. The greatest difficulty is the type and above all the magnitude of the crimes that were committed: there are no known laws, national or international, against murder on the massive scale on which it occurred. There is no punishment in legal tradition that will fit the crime of mass murder of several hundred thousand people committed to firing squads or gas chambers without even the semblance of a judicial process but simply by administrative action, because they belonged to a particular race or ethnic group. Whatever may be the accepted penalty for murder, whether it is death or lifelong imprisonment, it is hard to make a differentiation between those whose victims number a single person or tens of thousands. There is no special accepted penalty for mass murder and the matter is left to the discretion of the courts. There has been the criticism that they have not always found an acceptable answer, that in fact they have been too lenient. Yet two problems that might have presented difficulties were overcome early, as a result of the Nuremburg trial of war criminals held by the Allies in 1946: that those who had conducted or helped the administrative process required for sending millions to their death were guilty, whether they had directly participated in the killings or not; and that carrying out orders from a higher authority did not exempt a person from guilt.

Though the Nuremberg trial may have set the right standards for the treatment of war criminals, in the main it dealt only with the top leadership in the Nazi hierarchy. It did not help the West Germans, once the Federal Republic had been established, that during the period of military government, that is from 1945–49, the occupying powers had been responsible for the arrest and bringing to justice of war criminals and Nazi leaders and had used this responsibility in varying degrees, but in most cases in a half-hearted and haphazard manner. Generally speaking, the western occupying powers were baffled by what to do with war criminals, grappling with a problem that was hitherto unknown to them. Some British officials are known to have been doubtful about the expediency of bringing war criminals to justice, others were totally opposed to it. There seems no doubt that as a result of inconsistent application of justice to war criminals many were allowed to escape with too light sentences or were not brought to justice at all. The reproach that is often heard is that it was those in high positions and with direct responsibility for crimes who escaped justice, while the small-fry had to pay the penalty.

It is right here to draw the distinction between the former active Nazis who, if they were apprehended, were tried by denazification courts, and those who had actually committed crimes by killing or torturing human beings. The object of the denazification procedure was to identify those people who were known to have strong Nazi sympathies and to make sure that they were removed from positions of authority or leadership in the Federal Republic. They included civil servants or other government officers, teachers and, of course, judges or other persons in high legal positions. This was a logical part of the process of eradicating Nazism

74

from the country, dictated by the Potsdam Agreement of 1945, and necessary if the new Germany was to become a democratic state and any revival of Nazism was to be prevented. These people, so it was intended, were to be removed from public life but they were not normally accused of criminal acts. That their removal was not pursued with enough vigour or consistency is in no small measure due to the difficulty of replacing them as workers or professional men and women in a country whose population was decimated by the war; there was a great shortage of suitable candidates for filling the posts which should have been vacated by former active Nazis. The result was that many retained their posts of responsibility, and some do to this day. But the inadequacy of the denazification process also contributed to the more serious problem of war criminals: it did not help their conviction when judges and juries have sometimes been reluctant to commit men or women to long periods of imprisonment for crimes committed on behalf of an estbalishment of which they had themselves been active supporters, even if they did not harbour any residual sympathies for this establishment.

The hunting down of war criminals and the production of evidence presented further difficulties to the German authorities. Some of the worst offenders had left the country after the collapse of Hitler's empire. Many of the top leaders, including it was believed members of the Nazi government, had taken the precaution of depositing large sums of money abroad, in most cases with Swiss banks, which allowed those who escaped to survive financially and aid other fugitives who were not fortunate enough to have sufficient funds outside Germany. German citizens who were sought by various countries, including the Federal Republic, for having committed war crimes in Nazi-occupied Europe, were known to be living in some Middle Eastern countries and in Latin America, where they were protected by regimes which sympathised with their philosophy of life or, as in the case of Egypt, were prepared to use them in their own fight against the State of Israel. But even within the Federal Republic itself it was often difficult to lay hands on war criminals. Some had become outwardly respectable citizens, leading a lawful existence in the new Germany while their past record was either unknown to the authorities or had somehow escaped the dragnet of West German justice.

Even when sufficient evidence was available to lead to an arrest there were the difficulties of finding witnesses. The vast majority of the victims of Hitler's extermination policies were dead. Those who had survived were often unknown to the authorities or untraceable. Those who could be found were not always willing to testify, preferring rather to forget the horrors they had once been through and unable to face the ordeal of a lengthy trial that would bring back memories of their former experiences.

As time passed, the victims, often old or sick, could no longer recall the events with sufficient accuracy for their evidence to lead to a conviction. In 1958 the Standing Conference of Ministers of Justice[2] in the Federal Republic decided to set up the Central Office for the Pursuit of Nazi Crimes Committed outside Germany[3] under the direction of the Public Prosecutor of Ulm, Erwin Schüle, at Ludwigsburg. Its function was to speed up the process of bringing war criminals to trial.

Its task was not made easier by the fact that many documents had been removed by the Allied occupation authorities after the end of the Second World War and taken to various documentation centres all over the world. Most of these were not returned until 1967. Many war criminals were moreover known to the authorities of the Soviet Bloc countries, including the German Democratic Republic, on whose territories they had committed crimes. But in the 1950s and 1960s their governments were often less than helpful in producing the evidence which they possessed and which, had it been placed at the disposal of the legal authorities of the Federal Republic, would have facilitated the prosecution of many of the offenders. The state of mutual relations was mainly responsible for this: there were no diplomatic relations, for example, between the Federal Republic and all Soviet Bloc countries except the USSR until the early 1970s. The greatest obstacle was the German Democratic Republic with which relations, as has been emphasised, were particularly difficult. The GDR had masses of evidence and held many official documents dating from the period of the Third Reich and its activities in eastern Europe. But the East Berlin government made it a condition that the release of these should become the subject of negotiations between the two German governments. This Bonn would not agree to as it would have implied recognition of the German Democratic Republic. Here, as in other instances, intra-German politics became an obstacle to the solution of the Federal Republic's political problems. Meanwhile the East Berlin government exploited the West German difficulties for propaganda purposes and at times published evidence embarrassing to the West German authorities.

The Israeli government had a particular interest in seeing that war criminals, especially those who had participated in what Hitler had called 'the final solution of the Jewish problem', were brought to justice. To the State of Israel, seen by its people as representing the Jews all over the world, it was not only a moral issue that justice was done and that the torturers received due punishment for their crimes: there was also a deep concern that war criminals left at large or any active and unrepentant Nazis remaining in positions of power or influence should form a nucleus for the revival of Hitlerism and perhaps one day a new holocaust. It is in

76

that light that Israeli reactions to any upsurge of anti-Semitism anywhere, including the anti-Semitic eruptions in West Germany in 1959–60, should be seen. Israel is particularly anxious that the whole extent of the horrors of Hitler's final solution should be brought to light as a warning to future generations, not least the younger generation growing up within its own borders, of what may happen when a whole nation delivers itself up to the demons of an evil and inhuman ideology. That explains the strenuous efforts that were made and the risks that were taken in the face of international propriety to find and kidnap Adolf Eichmann, the most important administrator of mass murder, and the way in which he was tried. The Israelis measured the redemption of the German people not only by its willingness to make material amends but also by its efforts to rid public offices in Germany of former active Nazis and mete out adequate justice to those who had committed crimes against the Jewish people.

But not only the Israelis were interested in bringing war criminals to justice. There was great concern in the USA, a country with a large and influential Jewish population. Above, all European countries which had suffered from the German occupation during the War, where the SS and the Gestapo had been active, began to show an interest when it became clear that the Statute of Limitations would begin to operate soon in the Federal Republic, making further conviction of war criminals impossible. The countries affected by war crimes included in particular Denmark, Norway, the Low Countries, France, Italy and Yugoslavia. The worst sufferers had, however, been the eastern European countries where lack of co-operation by the authorities often hampered proceedings. Although they had a cardinal interest in having war criminals who operated on their territories brought to justice, their governments were not particularly anxious to aid the Federal Republic in overcoming its difficulties, especially when these difficulties helped to sour West German relations with the rest of the world. The West, for example, like Israel, was beginning to see the treatment of war criminals in the Federal Republic as a test of the West Germans' sincerity in their efforts to overcome the past, efforts which the German Democratic Republic consistently maintained had failed.

It is not surprising, therefore, that world opinion reacted angrily when it became known in 1964 that the Statute of Limitations might not be extended for war criminals beyond 9 May 1965 and that after that date no war criminal who had not been identified as such by that time could be brought to justice. Under West German law no person could be put on trial for murder unless legal proceedings had been initiated within 20 years of the alleged crime having been committed. Since the type of crime in question was not investigated as long as the Hitler regime was in power, being at that time not considered a crime at all, the 20-year period began on 8 May 1945, the day that the Third Reich surrendered to the Allies. The West German government, put on the defensive by protests from many quarters, especially from the United States, Israel and west European countries, including the Federal Republic, claimed that war crimes trials would continue because a large number of offenders were already known to the authorities but had, for one reason or another, not yet been brought before a court. To these the Statute of Limitations would not apply. This did not satisfy the critics. In view of the inadequacy of procedures against war criminals in the past and the slowness with which offenders had been discovered, it was suspected that many were still living freely in the Federal Republic or elsewhere and that it would take the authorities many more years to complete the process. Documentation centres in both the West and the East let it be known that they still had much untapped evidence which would be placed at the German authorities' disposal if and when they asked for it.

As the date approached when the Statute of Limitations would make it impossible for newly discovered war criminals to be tried, international concern increased. It was recalled that five years earlier the period for trying cases of manslaughter, robbery and torture had expired and had not been extended despite pleas that it should be. On that occasion the West German Minister of Justice, Fritz Schäffer, was able to say that the vast majority of crimes were not manslaughter but murder and that while most manslaughter cases were already before the courts, the search for murder cases would continue for another five years. On a visit to London in March 1964 Golda Meir said at a press conference that Israel was watching current trials of war criminals in the Federal Republic attentively. The Statute of Limitations should be extended. This was not a matter of revenge but was based on the principle that unconvicted war criminals should not feel free to live as if nothing had happened. It was not a judgment of the past but a guarantee for the future.[4] Commenting on a statement by Chancellor Erhard that the 20-year Statute of Limitations for murder was unlikely to be changed, a Foreign Ministry spokesman in Jerusalem said he hoped that the Bonn government would find ways to punish war criminals whenever they could be reached. As criticism from abroad began to build up, however, Erhard seemed to have changed his mind when at a press conference on 25 September 1964 he envisaged an extension of the Statute of Limitations for all cases of murder. As regards war crimes he told a journalist: 'For me it would be intolerable if such brutal and dastardly mass murders could no longer be subjected to punishment.'[5] The Federal Government also became concerned that the German Democratic Republic might publish documents after 8 May 1965 which incriminated West German citizens who could no longer be tried. In a statement on 15 October 1964 Erhard expressed the hope that a way might be found to exclude Nazi crimes from the Statute of Limitations.

This view was not shared by all Germans and certainly not by all members of Erhard's cabinet. War crimes trials were not popular in the Federal Republic. This was in part due to the apparent unevenness and unfairness with which they were thought to have been conducted. But there was a large body of opinion among the public which felt that the pursuit of war criminals could not go on for ever and that it was time to 'draw a line under the past'. It was indeed a shock to many to find men or women who had apparently lived normal, innocent and respectable lives for many years being suddenly accused of having actively participated in the tortures and mass murders during the war. Some Germans felt sympathy with the pleas made by most of the accused that they were under military orders at the time and could not have acted otherwise. But it was precisely the wish to 'draw the line under the past' that was widely criticised as a feckless form of escape from moral responsibility.

The members of the West German cabinet who were sensitive to world reactions nevertheless could not ignore the pressures of public opinion, especially in an election year.[6] Besides political reasons, moreover, there were legal objections to altering the Statute of Limitations: the Bonn Constitution states in Article 103 paragraph 2 that 'a deed can only be punished if it was a punishable deed before it was committed'. As in the case of the proposals for legislation against the West German scientists in Egypt, therefore, the legislators were faced with having to change the law in order to embrace acts that had already been committed. In any democratic society governed by the rule of law, statutes which have retroactive force and may have been passed to apply specifically to a group of people who committed certain acts *a posteriori* regarded as wrongful are clearly undesirable and possibly dangerous. In view of the vagaries of legal procedures in the Third Reich the Germans today might be expected to be particularly sensitive to such legislation. The many pleas that genocide did not come under the conventional categories of crime and that crimes against humanity came under international law and were exempt from Statutes of Limitations did not convince most legal experts in the Federal Republic. A fierce debate now began among the West German public and in the Federal Parliament.

There were three possible ways of dealing with the Statute of Limitations: to abolish it completely, to extend its time or to leave it to expire in May 1965. In view of the criticism against allowing it to expire and the difficulty of abolishing it altogether, an extension by four years to 1969 seemed to be the easiest way out. It would have meant that the 20-year period would be deemed to have started not at the end of the war but in 1949 when the Federal Republic was founded. This could be defended with the argument that West German courts under military government could not function as freely and independently as they could once the Federal Republic with it democratic system had been established. Such a four-year extension would require a reinterpretation of an existing law rather than the passing of a new one. This method received a great deal of support, including that of Dr Adenauer who had recently handed over the chancellorship to Erhard. Adenauer was opposed to complete abolition. The weakness of this proposal was that the four-year extension was too short and there were many who argued that even 1969 was too early for all war criminals to become known to the West German authorities. Yet another possibility, that while the Statute of Limitations should remain unchanged 'crimes against humanity' should be excluded from it was never seriously entertained. But on 11 November 1964 it was announced, despite Erhard's previously expressed views, that the Statute of Limitations would not be extended after all. The reason given was that such an extension would conflict with the constitution.

The announcement came at a press conference at which the Federal Government spokesman, von Hase, stated that 'after long discussion' the Federal Government had

> come to the conclusion that any extension with retrospective effect of the time limit in the Statute of Limitations for [the prosecution of] crimes is precluded by Article 103 of the Basic Law, however desirable such an extension might be for other reasons.

Von Hase then quoted the Article in question and added:

> This is the well known principle of *nulla poena sine lege*. It was the aim of the Federal Government to prevent at all costs that this very important question should lead to a controversial debate in parliament. The Federal Minister of Justice had therefore tried to discuss the matter with the parliamentary groups with a view to announcing the Federal Government's attitude in parliament and supporting it at the same time by a declaration of the three parties. This possibility was frustrated by yesterday's discussion in the Council of the Elders [of the Bundestag].[7]

These words barely conceal the amount of dissension that existed both in the ranks of the government and the political parties in the Bundestag. The SPD and a large number of members of the CDU/CSU group were not satisfied with this rather legalistic approach. Only the smallest party in the Bundestag, the Free Democrats, were united in their support for the maintenance of the existing position, but they were a constituent of the government coalition. One of their leaders, the Federal Minister of Justice Dr Ewalt Bucher, strongly defended the government's decision by upholding the primacy of the constitution in matters of this kind. But he also claimed, one wonders with how much conviction, that by the date of the expiry of the 20-year period a very substantial number of war criminals would be included in the dossiers of the judicial authorities, and, whether apprehended or not, would be tried by the courts because they would not be affected by the Statute of Limitations. Here von Hase's government statement quoted a report by the Minister of Justice saying that enquiries had been initiated against altogether 30,000 persons and up to 1 January 1964, 12,862 had been indicted because of Nazi crimes. Statistics published on 13 November 1964 by the Ministry of Justice indicated that from 1945 to the beginning of 1964 10,400 Nazi war criminals were sentenced by Allied and West German courts.[8] The statement added:

> ... the Federal Government intend to issue an appeal to world opinion ... asking for production of any evidence still existing which has not yet led to judicial action suspending the time limit. This attitude proves the political importance attached to the question by the Federal Government ...[9]

It is not clear whether this part of the statement was made to show goodwill or whether the government was really unaware of the masses of material that was still available and the complexities that this material could still create. But there was an immediate reaction from many parts of the world.

It was one of dismay and in some cases anger at the official announcement that the Statute of Limitations would not be extended. The Soviet Bloc countries were particularly vocal in their protest. The Soviet government, which had made some material available for the 'Auschwitz Trial'[10] held in the Federal Republic in 1963, but had also withheld documents at other times, denounced the West German decision as 'devoid of any political, legal or moral grounds'. Quoting Allied Declarations of 1942 and 1943, the Potsdam Agreement of 1945 and the UN General Assembly resolutions of 1946–47, all relating to the prosecution of war criminals, it said that all these declarations should be binding on the West German courts. It claimed that the Statute of Limitations, which existed in all countries, covered only 'ordinary criminal offences', not 'Hitlerite war crimes' as they were in contradiction with the standards of contemporary international law. Neither the Nuremburg War Crimes Trial nor Law Number 10 of the Allied Control Council in Germany had made provision for a Statute of Limitations. The statement quoted Article 25 of the West German Basic Law which said that 'the general standards of international law' ... have precedence over the laws'. 'West Germany's invocation of the 1871 Statute of Limitations' was 'contrary to present day international law' and 'a mockery of the memory of millions of people who were victims of the Nazis'.[11]

The East Germans joined the chorus. After reiterating the relevance of various Allied Declarations and parts of the West German Basic Law, *Neues Deutschland* claimed that 12,000 persons had been registered in Poland as having committed crimes on Polish territory; of these only 2,000 had been tried in the Federal Republic.[12] It also quoted the Chief Public Prosecutor of the Federal Republic, Dr Bauer, saying: 'There are still unknown cases, new denunciations are currently coming in ... an extension law of the Bundestag would be welcome' and a Professor of the Faculty of Justice of Prague University claiming that 27,000 war criminals remained untried.[13] Despite all these protests there were some positive moves by other communist governments; it was announced in Bonn on 6 January 1965 that Poland and Czechoslovakia had agreed to co-operate with the Central Office in Ludwigsburg in making previously unknown material available to help track down Nazi criminals. A party of West German legal experts would go to Warsaw within a month. A representative of the Czech Department of Justice was expected to arrive at Ludwigsburg with documents. The Bonn government welcomed the Polish and Czech attitude and hoped that other countries, especially the USSR, would follow their example.[14]

In Israel the reaction to the West German announcement was one of anger and disappointment. Relations between the Federal Republic and the Jewish State were at their lowest for some time owing to Bonn's failure to establish diplomatic relations and the presence of West German scientists and technicians in Egypt, against whom the Federal Government showed no sign of acting. After her statement in London in March 1964, the Foreign Minister Golda Meir had expressed her views on the question of German war criminals more strongly in the Knesset: 'Humanity cannot breathe freely so long as Nazi criminals are still at liberty. Such crimes cannot be measured by the same criteria as any others.'[15] Erhard's statement that the Statute of Limitations might be extended for all murderers had been severely criticised. The fortnightly *Israel Digest*, in a leading article, had stated that the implication that Nazi war criminals would be treated in the same way as others was received in Israel with astonishment if not bitterness, since 'it equated those responsible for liquidating six million Jews with other murderers'. Especially in Germany it was essential to prevent a situation where some of the worst

Nazi criminals who had so far evaded the charge of mass murder would be free by next May to come out of hiding and claim full privileges as citizens of the West German Republic.[16] 'Disappointment and indignation' was the reaction of the Deputy Prime Minister, Abba Eban, to the Bonn decision not to extend the 20-year period. Addressing the Knesset on 18 November 1964, he said that although the 'principle of prescription' was generally embodied in the legal systems of free regimes, the campaign of slaughter and torture perpetrated by the Nazi government was unique in the annals of criminality, as recognised at the Nuremberg War Crimes Trials and in the UN Genocide Act. The West German government should have shown sensitivity to the feelings of thousands of survivors of the Holocaust and to the principles of historic justice by separating the realm of general law from the 'special account that humanity must present to the Nazi criminals'.

> We take this stand not only in respect for the memory of the martyrs; it is even more categorically required in order to clear the conscience of our generation and safeguard the generations to come .... It is therefore the duty of the Israeli government to express profound disappointment and indignation at the statement made by the West German government on 11 November ....[17]

The world press generally added its criticism. A letter by Kurt A. Grossman in the *New York Times* said that West Germany could only become a deep-rooted democracy when it had purified itself of the poison in her blood stream. Estimates by the former US War Crimes prosecutor at Nuremberg, Robert W. Kempner, were that 10,000 murderers would benefit by the Statute of Limitations. 'They would be able to boast of their misdeeds without endangering their freedom.' The reason given for the decision 'not to follow the Nazi practice of making exceptions to legal principles to suit the government' was irrelevant when blatant murders of helpless people were committed without consideration of any legal or moral principles. The writer urged that the Statute of Limitations should be extended to 1969.[18] A letter signed simply 'Lawyer' appearing in *The Guardian* reiterated the view that Nazi crimes were crimes under international law. War crimes were defined by the 1945 Potsdam and London Agreements as well as by the Nuremberg judgments and Allied Control Council directives all of which also required that they should be punished; they did not lay down a time limit.[19] Both the West German government's failures and its difficulties were perhaps best summed up in a leading article of *The Guardian*: 'There is a timeless ring about the Potsdam paragraph demanding that "war criminals and those who participated in planning or carrying out the Nazi enterprise involving or resulting in atrocities or war crimes shall be arrested and brought to judgment"'. There was no Statute of Limitations in international law, which normally took precedence. This could have given the West Germans the chance to postpone the operation of the Statute. On the other hand, the Western Powers themselves – during the military regime – had overlooked war criminals, allowing them to return to positions of influence. One could not expect the Federal Government to pursue those whom the Allies had allowed to survive.[20]

The Bundestag debated the Federal Government decision not to extend the Statute of Limitations on 9 December and passed a motion which put the onus back on the Länder governments. They were asked to make a concerted effort to utilise all information that became available on war crimes through the Federal Government's worldwide appeal or from communist countries, and the Justice Minister, Dr Bucher, was instructed to report on progress by 1 March 1965. The possibility of an extension would then be re-examined. Only the FDP voted against the motion.[21] It seemed like a last desperate attempt to stave off what must have been in the minds of a majority of the Bundestag members: that an extension was inevitable. The Minister, a FDP member of the coalition cabinet, reiterated during the debate the successes already achieved in tracking down war criminals, especially after the government's worldwide appeal. In the course of the controversy which raged during the autumn months of 1964 he had developed all the arguments against any change and made it clear that he would resign if the Bundestag voted in favour of changes. At a press conference he went so far as to ask Jews to refrain from sending delegations or memoranda as the Federal Government could not be seen to yield to pressure. He added with surprising frankness: 'The public object; they do not want these trials.'[22] It may well have been his influence that at first swayed the cabinet against a decision to extend the Statute of Limitations.

Soon, however, increasing international pressure and the realisation that many more cases of war crimes would be brought to light but could not be ready by the deadline of 8 May 1965 led to a change of heart. On 8 March of that year 76 professors of law of West German universities appealed to the Bundestag to extend the Statute of Limitations, arguing that there were no constitutional objections to an extension for all murders and that failure to extend the Statute of Limitations would mean flouting the basic idea of the rule of law. The Secretary General of the United Nations, U Thant, said on 23 February that he was aware that nations of both eastern and western Europe were unanimous in demanding that legal prosecutions of Nazi crimes should continue.[23] A conference in London of resistance veterans and ex-soldiers of the Second World War from several countries made similar demands to the West German Ambassador on 9 March.[24] On 9 April in Geneva the UN Commission on Human Rights adopted a unanimous resolution urging all states 'to ensure that the criminals responsible for war crimes and crimes against humanity are traced, apprehended and equitably punished by competent courts'.[25] The report produced in early March by the Federal Minister of Justice following the Bundestag resolution of 9 December 1964 convinced the majority of the cabinet that the work of bringing war criminals to justice might have to continue after 8 May. The report stated that about 70,000 Germans had already been convicted of war and Nazi crimes and that proceedings against a further 13,000 were in the pipeline and would be instituted by the deadline. But new sources from eastern Europe had brought evidence of previously unknown criminal activities. Even with the further prosecution of Nazi crimes a complete uncovering of all the misdeeds and punishment befitting the personal responsibility of the perpetrators could not be guaranteed.[26] After it had considered the report the Federal Cabinet declared:

> Contrary to previous views the possibility cannot be excluded that after 8 May new facts will come to light which must give rise to new proceedings. The Federal Government will support the Bundestag in its efforts to create a possibility of satisfying justice while guaranteeing the principles of a constitutional state.[27]

During the debate on 10 March the Bundestag had before it the three proposals previously canvassed: a bill tabled by 50 CDU members, led by Berlin deputy Ernst Benda, to amend the Penal Code in order to extend the period of limitation for all crimes punishable by life imprisonment (which includes all murder) from 20 to 30 years; a bill tabled by the SPD to amend the Basic Law to cancel any limitation on prosecution for the perpetrators of murder or genocide; and Adenauer's proposal to move the commencement of the 20-year period from 8 May 1945 to 31 December 1949.[28] The Minister of Justice, Dr Bucher, remained adamant in his opposition, but admitted that some Nazi murderers not yet charged might escape. He said that a change 'under pressure from abroad' would do violence to the Basic Law, especially Article 102 which prohibits retroactive legislation. He added that no material had been received from the German Democratic Republic. He was supported by Dr Thomas Dehler, FDP, who insisted on a decision based on 'strict law'. Dr Rainer Barzel, CDU, on the other hand, supported Adenauer's proposal to extend the Statute of Limitations to 1969. SPD members speaking for complete abolition of the Statute of Limitations for murder or genocide took a moral line: the German people had a national responsibility to discharge, said Fritz Erler, as no one could deny that the perpetrators of Nazi crimes were Germans. Dr Adolf Arndt said that although not every German was legally guilty of crimes committed against the victims of Nazism the whole nation shared 'historic and moral guilt'. The Chancellor, Dr Ludwig Erhard, who, though basically in favour of changes had not exercised any great leadership in the matter, did not participate in the debate.[29] When the three proposals passed to the Legal Committee, it opted for the CDU proposal to extend the Statute of Limitations by another ten years to 30. But in a final compromise between the parties it was decided to withdraw both the CDU and the SPD proposal and on 25 March the Bundestag passed the Adenauer version of extending the Statute of Limitations to 31 December 1969 by 344 votes to 96.[30] The extension of course applied to all crimes of murder.

The exact reasons for the last-minute compromise are difficult to determine. According to the party leaders it was feared that neither the CDU nor the SPD proposals would have obtained a majority. That is possible if the FDP and a few members of the other parties would have voted against both notions. But it would still beg the question why there was such strong opposition to changes in the application of the Statute of Limitations when there were strong arguments in favour, the most plausible of which was that international law had been universally agreed as overriding the laws of individual states. This interpretation had been adopted by those other countries where a Statute of Limitations existed and the judiciary had to deal with war crimes. Once again – as in the case of German scientists in Egypt – the legalistic approach which hides behind the constitution seems unconvincing, although it could be said, again, that this approach could stem from a feeling of insecurity resulting from past mismanagement of democracy and abuse of the legal system under a totalitarian regime. The most likely reason for the opposition to a change in the law concerning the Statute of Limitations seems to be a deep-rooted aversion to the war crimes trials and a desire to end them as soon as possible. Not that there was any condoning of war crimes, except by a tiny minority. But can it be doubted that many people experienced a feeling of unease when someone was sent for trial for acts committed many years ago under a regime which they themselves had supported, especially when the accused claimed that he was under orders, which may have come from Hitler himself, to whose commands the others, including those who were now sitting in judgement, had themselves been subjected? After all, a large number of persons in prominent positions have been revealed as being former Nazi activists, members of the Nazi party or as having been involved in some official activity in the Third Reich. They include a Federal President, a Chancellor, several State Secretaries, judges, public prosecutors and others in influential positions in the Federal Republic. Some still hold high office. It cannot be surprising that these people, as individuals rather than a pressure group, formed an influential body of opinion which desired that the war crimes trials should end as soon as possible. Thus the argument that the impending parliamentary election was a factor for Bundestag members gains credence. At the time of the vote the election was just over six months away and it is likely that few of the members wanted to be identified with legislation that would lead to a long continuation of the unpopular trials, even if the compromise was unsatisfactory because it was too short.

It was indeed only a postponement of the problem until 1969 and was followed by a further extension until 1979 when the Statute of Limitations for all murder cases was abolished. In Israel, the country most emotionally involved in the affair, the Bundestag vote was also seen as a disappointing half-measure but it nevertheless helped to bring about its normalisation with the Federal Republic.

One episode concerning war criminals must be mentioned briefly here: the Eichmann trial. It had nothing to do with the 'overcoming of the past' by the German people, since both the capture of Eichmann and his trial were carried out by Israel. But it temporarily affected German–Israeli relations, even if only slightly. Eichmann had been one of the chief administrators of Hitler's 'final solution of the Jewish question' during the Second World War and was thus one of those who, under Hitler, had been responsible for the extermination of millions of people. After the war he had made his escape from Germany to South America, where he was arrested by the Israeli secret service and brought in total secrecy to Jerusalem in May 1960. His trial there lasted from April to December 1961. During the pre-trial period there was deep concern in the Federal Republic that the inevitable revelations and the publicity that would follow would bring the German people, whose good name had only recently been re-established, again into disrepute. Adenauer said at a press conference on 10 March 1961:

> Of course the Eichmann trial worries me, but not only the trial as such. Eichmann will get what he deserves. I have complete faith in the Israeli administration of justice. But I am concerned about the effect of what will be said on the way we Germans are judged generally.[31]

The main debate in Germany centred on the concern of the government and public figures not so much about what might be revealed at the trial but rather that it should be clear that the situation in Germany had changed since the Hitler period. To put it in different words: it was hoped that the world would recognise that what was on trial was the Hitler regime and not the Germans of today.

In this respect the Israeli government came to West Germany's aid. Prime Minister Ben-Gurion said in August 1961 in an interview with a German newspaper: 'My views about present day Germany have not changed. There is no longer a Nazi Germany. On the Israeli side there is readiness for close and normal relations and full cooperation.'[32] It is said that Ben-Gurion saw a specific aim in the trial: to remind young Israelis who had had no personal experience of the Holocaust, of the terrible crimes that had been committed against their people, lest they should become complacent, and to impress upon them that the State of Israel offered some protection against a recurrence of such crimes. But he also had to make sure that the new beginning in German–Israeli relations which he had helped to create should not be spoilt. It is generally assumed that because of his intervention a number of important public personalities in West Germany who could have provided evidence were not called to do so at the trial. One of these was Hans Globke, at one time State Secretary in the Chancellor's Office under Adenauer, who had to resign because of his associations with the Hitler government. In a telephone interview with a correspondent of the *International Herald Tribune*, F.J. Strauss, the West German defence minister during the Adenauer era, who had negotiated the secret arms deliveries to Israel, admitted that there had been links between the concessions made by the Israelis over who should or should not testify at the Eichmann trial and the delivery of certain weapons by Bonn to Israel, in the sense that 'the German side, with the agreement of Adenauer, fulfilled certain wishes expressed by the Israeli side'.[33]

In Israel the trial nevertheless stirred up new anti-German feeling among the population. The pressure of public opinion following the end of the trial[34] was strong enough to force the government to curtail private and cultural contacts between Israelis and West Germans. The move affected Israeli visits to the Federal Republic rather more than West German visits to Israel. In January 1962 a motion by Herut, the right-wing opposition party, to discontinue all cultural and educational relations with Germany was rejected by the Knesset. But Abba Eban, then Minister of Education and Culture, said in the debate that '... in view of the historical circumstances there should be special supervision over contacts with Germany, which should not be treated like other international contacts'. As a result travel by Israelis, especially young people, to West Germany for study, educational activities, entertainment or on invitation by West German organisations became subject to government permission and were generally restricted.[35] In a sense these regulations were a retrograde step, for contacts between the two countries had increased considerably in the previous few years. They caused much offence in the Federal Republic, especially among Israel's friends. But Eban, in a press interview, justified them as inevitable after the Eichmann trial, 'for memories that were abstract suddenly became vivid and concrete, particularly among the young who had not experienced the Holocaust'.[36] The restrictions were totally rescinded only after the Six Day War in 1967. But in the meantime economic and political relations between the two governments were not affected.

NOTES

1. The extermination of Jews was deliberate government policy, carried out on the personal orders of Hitler. Obeying orders of superiors was not accepted at the Nuremberg War Crimes Trial in 1945–46 as extenuating circumstances in war crimes.
2. The administration of justice is the function of the Länder in the Federal Republic.
3. Zentralstelle für die Verfolgung im Ausland begangener NS – Mordverbrechen.
4. *Neue Zürcher Zeitung*, 7 March 1964.
5. *Deutschkron*, op. cit., p. 261.
6. Bundestag elections were due in October 1965.
7. *Keesing's Contemporary Archives*, p. 20511A (1965–66).
8. Ibid.
9. Ibid.
10. Trial of members of camp personnel 20 Dec. 1963–20 Aug. 1965. in Frankfurt-am-Main.
11. *Keesing's Contemporary Archives*, p. 20532A (1965–66).
12. *Neues Deutschland*, 12 Nov. 1964.
13. Ibid., 17 Nov. 1964.
14. *Keesing's Contemporary Archives*, p. 20532A.
15. Statement in Knesset on 20 May 1964 quoted by *Israel Digest*, 23 Oct. 1964.
16. *Israel Digest*, 23 Oct. 1964.
17. *Keesing's Contemporary Archives*, p. 20511A (1964–65).
18. *New York Times*, 20 Nov. 1964.
19. *The Guardian*, 27 Nov. 1964.
20. *The Guardian*, 14 Dec. 1964.
21. *Keesing's Contemporary Archives*, p. 20511A (1964–65).
22. *Deutschkron*, op. cit., p. 266.
23. *Keesing's Contemporary Archives*, p. 20951A (1964–65).
24. Ibid.
25. Ibid.
26. Ibid.
27. Ibid.
28. Ibid.
29. Ibid.
30. Ibid.
31. Deligdisch, op. cit., p. 66.
32. Deligdisch, op. cit., p. 69.
33. *International Herald Tribune*, 9 March 1965.
34. Eichmann was found guilty on 15 Dec. 1961 and sentenced to death. The verdict was generally approved of in Germany.
35. *Israel Digest*, vol. V, no. 2, 19 Jan. 1962.
36. *New York Times*, 14 Jan. 1962.

7

1965: Year of Crisis and Crisis Resolution

By late 1964 relations between the Federal Republic and Israel had reached their lowest ebb since they began in the early 1950s. The irritation, it must be said, was nearly all on Israel's side. For nine years the West German government had refused to establish diplomatic relations when it had an opportunity to do so. While not making much difference to the conduct of day-to-day economic and political business between the two states, it was resented as a slight and condemned by the Israeli public and media as discrimination, even if the government, for good reasons, had to keep its cool. Then there was the unresolved dispute over the West German scientists and technicians working for the Egyptian government on armaments, whom the Federal Government had not been able to call home. Finally, there was ill-feeling over the prosecution of war criminals, leading early in 1965 to the half-measure decreed by the Bundestag of extending by four and a half years the Statute of Limitations for all murderers. On the credit side, however, the West Germans were continuing to keep strictly to the restitution payments under the Luxembourg Agreement which was due to expire in 1966, and more aid was going to Israel as a result of the meeting between Adenauer and Ben-Gurion in 1960. But most important, the secret flow of weapons was continuing from the Federal Republic to Israel and was, in fact, being increased. These latter factors, restitution, economic aid and above all military supplies were undoubtedly the reasons why the Israeli government did not join the growing tide of protests that came from the media, but confined itself to occasional criticism, subdued even in the emotive case of the West German scientists. Then in October 1964 the secret of the West German arms supplies was revealed to the world, though not necessarily to the top politicians, who may well have known for some time. But the 'revelation' provided the Egyptian President with a reason to challenge the Hallstein Doctrine. The crisis ended with a humiliation for the Federal Government. It stopped the arms flow but also led to the establishment of diplomatic relations between the Federal Republic and Israel, and to

0

more economic aid for the latter. The other two problems, the scientists in Egypt and the differences over the Statute of Limitations faded away. Relations now became officially normal – the question was whether they actually improved.

Looked at from the West German side, the Federal Government's policies regarding Israel in 1964 were strangely inconsistent, reflecting the dilemma of the Federal Republic over past German treatment of the Jews on the one hand and the necessity of protecting the national interest in a divided and unsettled world on the other. Thus the Federal Republic was continuing to pay large sums of restitution in fulfilment of the Luxembourg Agreement for the mass killings of Jews and the confiscation of their property under the Third Reich, but was denying the Jewish State the elementary courtesy of recognition as a sovereign state in international law for fear that this might lead the Arab states to recognise the German Democratic Republic, thereby hindering the prospects of future German reunification in the eyes of West German politicians. In the military field the Federal Republic was secretly delivering sizeable quantities of modern weapons completely free of charge to Israel in order to help it build up its defences against its Arab neighbours who threatened to destroy it and who had obtained massive amounts of modern arms for this purpose from the Soviet Bloc; yet at the same time the Federal Government seemed unable to stop some of its own nationals aiding Egypt technologically to improve its weapons output, which was likely one day to be used against Israel. Again, that Bonn was guarding its neutrality in the Arab–Israeli conflict and trying to keep good relations with both sides is not at all unnatural. But the intensity with which these relations were pursued by the Federal Government was bound sooner or later to lead to a crisis. Added to all this was the vulnerability of the still relatively new West German state, one part of a divided Germany, a state which had recently emerged out of defeat and universal disgrace and which, with relatively little military power or political influence, was trying to hold its own in a complex, divided and potentially violent international system.

The crisis developed slowly at first after the *Frankfurter Rundschau* in October 1964 first reported the West German arms deliveries to Israel.[1] That the secret leaked out cannot occasion surprise, since so many people in several parts of the world now knew about it. Apart from a number of Bundestag deputies and Ministry of Defence officials, some of whom were known to have disapproved of the secret arms agreement, the arms deliveries were known to officers of the Bundeswehr, some of whom periodically travelled to Israel in the course of their duties. In Israel, too, they were known to the government, the Leader of the Opposition and the defence establishment. In addition, they were known to the Federal

Republic's allies, some of whom, especially Italy, provided transit facilities for arms consignments between West Germany and Israel. There were also the numerous oblique references to 'unmentionable' goings-on between the Federal Republic and Israel made by Strauss and several members of the Israeli government. President Nasser of Egypt claimed that he had known about regular West German arms shipments to Israel since 1963; there is no reason to disbelieve that claim. If he took no action it could be that he considered the shipments insignificant or, more likely, that he was waiting for an opportune moment to use them, possibly as a bargaining counter. This, and the fact that after years of silence the details appeared suddenly in several parts of the world suggest that they were deliberately leaked by someone.

Neither the Israeli nor the Federal Government at first admitted that there was a secret arms agreement. The former dismissed it as propaganda emanating from Cairo. The official spokesman of the Federal Government, without actually denying the arms deliveries, made the error of saying that there was indeed scientific co-operation between the Federal Republic and Israel for peaceful purposes, that is in the nuclear and biological fields; several West German scientists were currently engaged in research at the Weizmann Institute.[2] This statement was bound to renew speculation in East Germany and some Arab countries, denied by Israel, that there was scientific co-operation in the nuclear field for military purposes. The West German statement could also, of course, weaken the Israeli case against the West German scientists in Egypt. It was as if it had the effect of confirming rather than denying the existence of an arms agreement, for within days the western press printed correct if incomplete information about the weapons supplied and the conditions under which it was possible to send them without members of the West German government and parliament knowing about it.

Several days after the first statement, the West German government was therefore obliged to admit that military equipment was being sent to Israel and training given to Israeli soldiers 'following talks between the former Chancellor Adenauer and the former chief of the Israeli government, Ben-Gurion',[3] though not that there was an agreement. The spokesman also accepted that this would lead to Arab protests. But concern was already growing in Israeli government circles about what the effect of these protests might be on the Federal Government. According to Shinnar, rumours began to circulate after the revelation of the secret that the Federal Government, fearing Arab reactions, might stop all arms exports to areas of political conflict, such as the Middle East, as had in fact been official West German policy. The Israeli representative was therefore called to Jerusalem for consultations and very soon returned to Bonn with an oral message from Prime Minister Levi Eshkol to Chancellor Erhard. In it the Israeli head of government approved the West German admission of the arms deliveries but warned that an arms embargo would give the impression that the Federal Government was allowing itself to be unduly pressurised by the Arab states. It would also, in the words of Eshkol, 'provoke a comparison with the efforts[4] of the Federal Government, so far in vain, to put an end to the activities of the German scientists in Egypt'.[5] This message may have prevented an early announcement by the Federal Government of a stoppage of the West German arms flow to Israel which, as Shinnar comments, would have been seen in Israel as a capitulation by the Federal Republic in the face of the Arab threats and a sacrifice of Israel's vital interests; this would in turn have led to unforeseeable reactions by the world press and grave consequences for the future of the already strained relations between the two countries.[6]

It should be mentioned here that Eshkol continued his predecessor's policy towards the Federal Republic but without the same urgency that Ben-Gurion had put into his relations with the Germans and without sharing Ben-Gurion's desire for reconciliation, which formed the basis of his attitude towards the new, post-war Germany. Eshkol, though cautious, was much more critical of what he considered the Federal Government's poor showing on the questions of diplomatic relations, the scientists in Egypt and the Statute of Limitations. Speaking in the Knesset on 12 October 1964 he noted that Erhard and former Chancellor Adenauer had condemned the scientists' activities as had many West German public figures, but nothing had changed. He did, however, add that tension between Israel and the new Germany would not be in Israel's interest and referred – in a similar vein to his predecessor – to 'the forces of renewal in Germany'.[7] Erhard politely replied to this statement three days later, but beyond regrets and disapproval of the scientists he had nothing new to offer. A small consolation came a week later from a Federal Government spokesman who said that several West German experts working in the Egyptian armaments industry were about to leave. Some would leave by the end of the year, others, including Kleinwächter, had already gone.[8]

In the closing weeks of 1964 there seemed little prospect of these irritants being removed, as the question of the Statute of Limitations, too, was no nearer a solution. That the Federal Government now came under pressure over its relations with the Arab states following the revelation of the secret arms agreement did not help. The Erhard cabinet at this moment was unlikely to want to offend the Arabs by establishing diplomatic relations with Israel – though the question of opening a 'mission' in Tel Aviv had come up again for discussion – or by taking any drastic action to remove the scientists and technicians still remaining in Egypt.

Meanwhile Eshkol announced that he would meet Erhard at some unspecified time 'on neutral ground'. On the other hand King Hussein of Jordan was expected in Bonn, and there was a standing invitation to President Nasser to visit the Federal Republic. This kind of summitry seems to have given some hope to the West German government of finding a way of extricating itself from a very difficult situation. But meanwhile the arms deliveries to Israel continued if anything at an accelerated pace in case they might have to be stopped altogether, while the Arab League threatened that its members would sever diplomatic relations with Bonn and might recognise the GDR if the arms flow to Israel did not cease.

In November 1964 Dr Eugen Gerstenmaier, President of the West German Bundestag, visited Cairo at the invitation of his Egyptian counterpart, the Speaker of the Parliament of the United Arab Republic. The visit had been planned several months before the crisis had started. It now had a special significance, especially since Gerstenmaier was known for his pro-Israeli sympathies, had recently paid a visit to Israel and had gone on record as advocating the early establishment of diplomatic relations between the two countries. The discussion on establishing a West German 'mission' in Tel Aviv, the equivalent of the Israeli mission in Cologne, constituted a first attempt to reopen this question since 1957. The Federal Government, having accepted without enthusiasm the establishment of East German consulates-general in Egypt and Syria, had felt that Arab leaders could hardly object to what was only an unofficial office which implied no recognition of Israel in international law. Prime Minister Eshkol, however, would hear of nothing short of full diplomatic relations, having previously criticised the West German government for failing to offer to establish them. It was believed that among Gerstenmaier's tasks had been to sound out the Egyptian government's attitude to a possible recognition of Israel by the Federal Republic and to raise the question of the West German scientists and technicians still employed in the Egyptian armaments industry. At the same time, as a sweetener, he was offering more economic aid. He was also renewing the invitation for Nasser to visit Bonn.

Gerstenmaier had apparently read the reports of West German arms deliveries to Israel with incredulity.[9] When he was received by President Nasser, who gave him chapter and verse and produced convincing evidence, his reaction was one of shock and anger. No request for authorisation of the arms deliveries had been made to the Bundestag and he had not been one of the small number of deputies who had been informed. It may be assumed that if he had been he, like many other West German parliamentarians, would have disapproved not only of the arms

94

deliveries per se, but also of the method adopted to carry them out. But to be faced with the facts by the Egyptian President for the first time was an extreme embarrassment. It is difficult to see why the Chancellor should have taken such a risk by not briefing his emissary. Erhard did indeed know of the agreement with Israel, having simply taken it over when Adenauer retired and continued to authorise the arms deliveries.

Despite Nasser's reproaches Gerstenmaier stated on his return to Bonn that his interlocuteur had not been unreasonable. Over the question of West German diplomatic relations with Israel he seems to have gained the impression that Nasser would not raise serious objections, provided the flow of weapons from the Federal Republic to Israel ceased. But when he conveyed this to the Chancellor no decision was taken and nothing changed. The cabinet was divided on the issue: Foreign Minister Gerhard Schröder was in favour of stopping the arms deliveries immediately and of adopting in practice the policy which the Federal Government had pursued in theory: that of not sending arms to 'areas of tension'. But there were voices both inside and outside the government strongly in favour of continuing the arms supplies as long as possible. These were reinforced by the fact that the arms agreement was supported by the US government which, as was stated earlier, was anxious to help Israel strengthen its defences because of the heavy supplies of arms received by some Arab states from the Soviet Bloc,[10] without being seen to supply Israel with weapons for fear of losing its standing with the pro-Western Arab states. The other factor which made it difficult for the Federal Government to end its military aid to Israel was the fear of Israeli reactions. With the increasing irritation in Israel, echoed by a large section of West German public opinion and the media, over diplomatic relations, the scientists in Egypt and the unwillingness to extend or abolish the Statute of Limitations, an arms embargo against Israel would have dealt a serious blow to the already strained West German–Israeli relations.

Erhard's utterances on the subject of relations with Israel indicate that he was as dedicated to reconciliation with the Jews and with the Jewish State as Adenauer had been. But he was a weaker personality and a less shrewd politician than his predecessor and it was his misfortune that by the mid-1960s the world political environment had changed and his country, through its foreign policy which began with Adenauer, had manoeuvred itself into a position between the warring Arabs and Israelis from which it could not extricate itself without offending either one side or the other. As one West German commentator put it acidly:

The Federal Government, in its treatment of Israel and the Arab states, had the pleasant hope of being able to pull all the fish ashore

95

with a net of tenuous or half-concealed relations. To be good friends with the Arabs, the arch-enemies of Israel, to concoct closer contacts with the Jewish State than can be recognised at first sight and thus to please the United States at the same time – that was the recipe.[11]

Because of the indecision of Erhard's government the great danger at this moment was that this situation could lead to a crisis as a result of which the Federal Republic might lose the goodwill of both sides. The Federal Government now looked for some support from its western friends, especially its major transatlantic ally, which had contributed to its current predicament. Not surprisingly perhaps, there appeared at first to be little response.

West German press comments were subdued at first and, like those of the politicians, showed bewilderment as the story of what had occurred unfolded. The Social Democrats expressed anger and let it be known that they had always been opposed to the sending of arms to areas of conflict. It was noted that the Federal Republic was sending arms to Israel and tolerating West German scientists contributing to Egypt's war machine. Herbert Wehner, a leading SPD deputy in the Bundestag, referred in a television broadcast to government policy towards Israel as 'official hypocrisy'. What were needed, he added, were clean diplomatic relations, not a game of hide-and-seek.[12] A commentator of the left-liberal *Frankfurter Rundschau* noted that 'Egypt and Israel were arming themselves with German help or with the participation of Germans against each other'.[13] That a secret arms agreement had been concluded came under special attack, since it was felt that such secrets cannot be preserved for long and could constitute a danger for the Federal Republic. Fear was expressed by some proponents of the Hallstein Doctrine that the revelation of the secret could lead to some Arab states recognising the German Democratic Republic, thus undermining West German reunification policy.

While these arguments were going on the Federal Government continued to do nothing. The failure of the government to act quickly and decisively by following the advice of the Foreign Minister to end all arms shipments to areas of tension, thereby effectively stopping military aid to Israel, was given by many West German critics as the reason for what happened next: the announcement in Cairo on 26 January 1965 that the Egyptian government had invited Walter Ulbricht, the First Secretary of the SED in the German Democratic Republic, on an official visit to the Egyptian capital. It was this which was to turn a West German embarrassment and a contretemps in the relations between the Federal Republic and the Arab world into a major diplomatic crisis. Nasser himself, when challenged by Bonn gave the arms deliveries to Israel as a justification for

Arab anger if not as a reason for the invitation and at the same time again threatened to recognise fully the GDR if they did not stop. But on his own admission he had known of them more than two years earlier and had not reacted until details were revealed in the world press, and even then he had waited three months before issuing the invitation. The real reasons for it must therefore be sought elsewhere rather than in West German–Israeli relations.

Several theories have been advanced by politicians and publicists trying to explain why Nasser invited the head of the East German state at that moment, when he must have known that this would adversely affect his relations with Bonn. He may not, of course, have realised that the Federal Government would react so strongly, since on the occasion of the visit in 1959 of Grotewohl, then Premier of the German Democratic Republic, the reaction was a mild one. The Egyptian government at that time, however, made it clear that this visit did not constitute recognition of the GDR and the West Germans had been satisfied. No such assurance accompanied the announcement of the invitation to Ulbricht.

The Egyptian President's reasons for challenging the Federal Republic and thereby indirectly the West in general must be sought in the context of the political and economic situation in which he found himself at the time, and indirectly, of developments taking place within the Soviet Bloc. Khrushchev's individual style of foreign policy had led to problems for the USSR and was severely criticised inside his own country. From Moscow's point of view the Cuban crisis had been a failure and humiliation. Soviet policy over Berlin in the late 1950s and early 1960s – an ultimatum, later allowed to lapse, then renewed and abandoned again without having succeeded in its aim, leading only to the rigid division of the city – and the proposals for a confederation of the two German states, also later abandoned, had not brought any advantages and had caused a measure of irritation in East Germany. Above all they had brought the international recognition of the GDR no nearer. In the early 1960s, therefore, the East Berlin government, feeling stronger because of economic recovery, began its own diplomatic efforts to gain influence in the Third World in the hope of breaking through the restrictions imposed on it by the Hallstein Doctrine. It was able for the first time to back up its diplomatic offensive by modest offers of economic aid. The unstable Middle East, where West German efforts had established a precarious balance of relationships with Israel and the Arab states, was fertile ground for diplomatic action.

The main reason for Nasser's Ulbricht coup, however, lies in the Egyptian President's precarious internal and external situation and his need to turn to the Soviets for help. In foreign affairs, especially in his relations with other Arab states, his government had encountered failure.

The union with Syria, through which Egypt had formed the UAR, had broken up in 1961 and Egypt's relations with that country were cool.[14] In the Yemeni civil war, in which Egypt supported the rebels with troops and weapons against the royalist regime supported by Saudi Arabia, the Egyptian forces were suffering reverses and the war, having reached stalemate, was draining the limited resources of the Egyptian economy. Nasser's failures in Yemen could seriously weaken his claim to Arab leadership, while their economic consequences were leading to food shortages at home. A lack of foreign currency was also threatening the success of his second five-year plan. But the USSR was in the midst of reappraisal following Khrushchev's dismissal in October 1964. Khrushchev had been generous in his aid to Egypt, despite Nasser's treatment of Egyptian communists. The changes in Moscow of October 1964 brought with them criticism of, among other matters, Khrushchev's extravagant 'rouble dip' in the Middle East. There were fears in Cairo that his promises of further Soviet aid to Egypt might not be honoured at a time when it was most needed. When, however, in December 1964 the Soviet deputy premier, Alexander Shelepin visited Cairo, he brought with him credits worth DM1.1 billion, but would not go beyond the promises of aid made by Khrushchev. It seems certain that he drew the Egyptian leader's attention to the possibility that the German Democratic Republic might be prepared to give further aid, especially if some kind of return favour could be granted.[15]

Whether full recognition was suggested or simply an invitation is not known; what is clear is that Ulbricht did offer DM320 million in credit.[16] Nevertheless, while the Egyptian government was in dire need of aid, yet unsure about the future attitude of the new men in charge in the Kremlin, an invitation to the chief executive of the East German state was the least that might have been expected from the Soviet side. In view of the furore this could create in Bonn, the secret West German–Israeli arms agreement provided a welcome pretext.

The interest of the GDR in the countries of the Third World was largely to enlist their support in the East Germans' struggle with the Federal Republic. Because they considered themselves 'neutral' in the East-West conflict, the non-aligned states might be persuaded to recognise both Germanies, thus accepting the 'two-state theory' upheld by the Soviet Union and undermining the Hallstein Doctrine. The USSR was not unhappy about the German Democratic Republic's progress in the Third World because it would counteract Chinese successes there.[17] Moscow may have also welcomed the opportunity to be of service to the East Berlin regime after recent disagreements with it over Berlin during the period of Khrushchev's rule. A blow against Bonn, moreover, would this time also be a blow against Israel, which both the Soviet Union and the German Democratic Republic regarded as a 'military bridgehead of the US' in the Middle East. Finally Moscow may have hoped to gain influence itself, through Nasser, in other Middle Eastern and African states such as Yemen and Congo, both deeply rent by civil war, and in which the Egyptian leader had become involved. Hence the increased arms deliveries to Egypt, which also followed the Shelepin visit.

Nasser did not, of course, admit that Soviet pressure had brought about the 'left turn' he seemed to world opinion to have taken. Such an admission would have indicated failure of his policies at home and abroad and given the impression that he was abandoning the policy of non-alignment in the East-West conflict.[18] It would also have been unpopular in Egypt itself, where Soviet domination was resented almost as much as excessive western influence. Non-alignment was the policy of most of the Arab world, of which he aspired to be the leader. Egyptian relations with the US were particularly bad at the time, despite the fact that Washington was giving economic aid, including much-needed foodstuffs. The burning of the American Library by a mob in Cairo, the shooting down of a private American plane and the supply of arms to the Congo rebels who were fighting a government supported by the United States did not stop this aid, despite objections from Congress, because the US feared that Egypt would move still further into the Soviet orbit. Now the conflict with Bonn seemed a further sign that Nasser was turning against the West. The advantage of using the West German–Israeli arms agreement as a pretext for the Ulbricht invitation was firstly that it did not directly relate to East-West relations and secondly that any move against Israel would elicit support from other Arab states. For this he did not have to wait long: at a conference of heads of governments of the Arab League states in Cairo in early January 1965 it was decided, undoubtedly with Egyptian inspiration, to 'adopt a unified plan in confronting any foreign power which might seek to establish new relations with Israel and to strengthen its aggressive efforts',[19] a threat clearly directed against Bonn. The Egyptian public had been prepared for this approach before the announcement of the Ulbricht visit by a press campaign against the Federal Republic over its relations with Israel and the revelation of the secret arms pact.[20]

Egyptian anger with the Federal Republic reached new heights early in February when the first West German reactions to the visit had been voiced. Muhammad Heikal, writing in *Al Ahram*, generally regarded as a mouthpiece of the Egyptian government, poured out a long list of reproaches against the Federal Government: the Federal Republic had committed a hostile act against the whole Arab world and dissipated a huge fund of Arab goodwill towards the Germans; the Arabs had always held the Germans in high esteem because of their 'pertinacity, precision, seriousness and firmness'; Egypt had borne no grudge when the Federal Republic concluded the Restitution Agreement with Israel, nor when it withdrew its offer of finance for the Aswan Dam; even the rumoured arms agreement, first thought to consist of a loan to Israel to buy arms, was not resented. 'No one then imagined,' wrote Heikal, 'that it was true.' The uproar over the West German scientists in Egypt, Heikal claimed, had been engineered to cover up something much more dangerous; an agreement to give Israel arms as a present. Heikal accused the Federal Republic of hypocrisy: it had sent Nasser an invitation, offered more aid for Egyptian industrial projects and made great play of its non-recognition of Israel. 'Recognition or non-recognition of Israel is of no value: what the Federal Republic did was worse than recognition.' Then came an argument that has frequently been expressed in Arab and especially in Palestinian circles: that the German people shouldered a special responsibility for the Palestine tragedy, since their persecution of the Jews had provided the prextext for the establishment of the Jewish State and a home in Palestine for the Jews who were persecuted and came out of concentration camps. 'Now Germany, which wants to make up for the sins of the Nazi regime, makes the Arabs pay with their security.'[21]

Some of these points were repeated by Nasser himself when he met the West German ambassador, Georg Federer, at the latter's request. Only one remark made by Nasser during this conversation is of interest here. When Federer made the point that the Ulbricht visit would affect West German economic aid to Egypt – meaning that it might be reduced or cut altogether – Nasser's reply was: 'We did not get aid from you or anyone else. You participated in industrial projects and we paid most of it at six per cent interest. That is not aid!'[22] If this remark was indeed made by the Egyptian President then he seems to have simply brushed aside the Federal Government's strongest card in enforcing the Hallstein Doctrine: its ability to cut off economic aid to developing countries if the Doctrine was infringed. Of whatever use the West German credits may have been to his country, Nasser appears to have been prepared to forego them, possibly hoping to obtain enough finance, perhaps even on better terms, elsewhere. Or he may have been calling what he thought was the Federal Government's bluff.

In the days following the Nasser–Federer dialogue the Federal authorities had still not officially admitted that there was an agreement to supply Israel with arms. But official West German anger at the Ulbricht invitation was expressed by the government spokesman, von Hase, when he said that the visit by Ulbricht, whom he called the 'arch-enemy of German unity', would cause the gravest changes in relations between the West German and Egyptian peoples and governments. The Federal Republic would reconsider the promise of more development aid unless the visit was cancelled.[23] The Federal Government was now in a serious dilemma. If it continued the flow of arms to Israel there was a real risk that Egypt, followed by other Arab states, would recognise the German Democratic Republic. The Hallstein Doctrine was now perceived to be in greater danger because of the arms agreement than over a possible West German recognition of Israel. Furthermore Egypt and other Arab states would most likely break off diplomatic relations with Bonn, or Bonn with them if they recognised the GDR, and that, in turn, would put an end to the Federal Republic's special position in the Middle East, a development which could only bring further benefits to the Soviet Bloc. If, on the other hand, the agreement was ended and the arms shipments were stopped, this would cause a serious conflict with Israel and it would be said by the Israelis and many of their sympathisers everywhere, including those in the Federal Republic, that the Erhard government had capitulated to Arab blackmail, that the Arabs were in fact determining West German foreign policy and would do so increasingly in the future. The Hallstein Doctrine would therefore still be at risk. So would the Federal Republic's policy of reconciliation with Israel.

It was with such arguments that pressure was increasingly applied on the government within the Federal Republic. The opposition SPD, for long a staunch supporter of West German economic aid for Israel, believed that the arms deliveries should cease. The deputy leader of the SPD in the Bundestag, Wischnewski, demanded an investigation into the military training and aid by the West German parliament. He believed that the Federal Government should renew its decision not to deliver arms to areas of tension, but it should also solve the problem of the scientists and technicians in Egypt by forcing them to come home.[24] On the other side, Adenauer, still chairman of the CDU, the senior partner of the government coalition, suggested that the government should react to 'Nasser's new attempt at blackmail' by a strict application of the Hallstein Doctrine, 'otherwise it would make us look ridiculous and we would lose respect among the other peoples'. He was supported by Professor Böhm, another CDU deputy and strong advocate of West German recognition of Israel, who said that by inviting Ulbricht, Cairo had already carried out the threat of action it would take if Bonn were to establish diplomatic relations with Israel and the way was therefore free to establish such relations.[25] The Federal Government, however, seems still to have been unable to make a decision other than to adopt a wait-and-see attitude, hoping that it might yet be able to persuade Nasser to cancel the invitation or at least to downgrade it to something much less formal. Meanwhile, according to von Hase, the Federal Government would announce no counter-measures until it had received a final assessment of all the facts. The West German ambassador would remain in Bonn and be at hand to advise on West German diplomatic actions towards the other Arab states.[26]

The first reaction of the West German press was one of dismay. Most newspapers linked the invitation of Ulbricht to Cairo with the arms deliveries to Israel. Some said they should have been stopped as soon as the secret arms deal had been revealed. Despite criticism of the Hallstein Doctrine in the past, the media on the whole showed concern that the whole structure of Bonn's foreign policy might now be in danger. The *Frankfurter Allgemeine*, however, called the invitation a 'declaration of war' by Cairo,[27] and asked for an end to West German development aid to countries which play host to the 'dividers of Germany'.[28] *Die Welt* said that the Federal Government should withdraw the invitation to Nasser if Ulbricht went to Cairo. A withdrawal of this invitation would be

> the least possible degree of self-respect which the government owes itself if it does not want ... in the eyes of the world to write off its claim to sole representation of all Germans. Bonn must make it clear to Nasser ... that his choice is not how to extract still greater concessions from the Federal Government, but to decide for or against Bonn.[29]

The liberal *Die Zeit*, after relating that the Federal Republic had granted Egypt DM1,200 million in aid, compared with about DM1,000 million granted by the USSR and DM313 million by the GDR, asked whether, given the economic misery and communist agitation in Egypt as well as the conflict with the USA, Nasser could afford to break with the Federal Republic and deliver himself into the hands of the USSR.[30] The conservative *Handelsblatt* commented:

> President Nasser, who deliberately offends us despite generous capital aid, cannot expect that we will continue to make allowances for him. The time has therefore come to take up the long overdue question of diplomatic relations with Israel.[31]

The *Rheinischer Merkur* (CDU) warned that the government would set off a dangerous chain reaction if it did not apply to Egypt the 'policy of breaking off diplomatic relations with countries that recognise the GDR'. The government should not let itself be intimidated by threats.[32] The independent left-wing *Neue Rheinzeitung* believed that Bonn now intended to show the Arab nations what a loss of West German aid could mean for them. It hoped that the Arab nations, aware of the importance of West German aid, would put Nasser under pressure.[33]

Up to this point the crisis seemed to stem from a conflict between the Federal Republic and Egypt or, in the sense that they had threatened to follow Nasser's line and recognise the GDR, between the Federal Republic and several Arab states. Israel, or rather the West German arms deliveries to that country, had played only a small part, although this was not generally recognised at the time. The Israelis were, however, becoming concerned that the pressure might become too much for the Federal Government to withstand. The situation changed when Federal Government officials first dropped hints that the country might no longer deliver arms to 'areas of tension'. Word first came on 8 February 1965 from the spokesman, von Hase, that the government proposed to submit legislation to the Bundestag to that effect. This was interpreted, correctly, to mean that the flow of weapons to Israel would be stopped, though this was not specifically stated. The spokesman added, however, that 70 per cent of the Israeli arms agreement had already been fulfilled. At the same time the Federal Government accepted an offer by Spain to mediate in the dispute with Egypt. The Spanish government had excellent relations with both the Federal Republic and all Arab states and Spain was one of the few countries that had not recognised Israel, from the Arab point of view a distinct advantage. The Marquess de Nerva, a Spanish Foreign Office official, travelled to Cairo to act as mediator. He must have taken with him a definite pledge that the Federal Government would end the arms aid to Israel, since a statement to this effect was made by the Egyptian Prime Minister, Ali Sabry, in the Egyptian National Assembly on 10 February. He claimed to have received this assurance from de Nerva. Chancellor Erhard confirmed it two days later, saying that the Federal Government would henceforth ban all arms deliveries to areas of tension and that the government was negotiating with Israel about 'other ways' of fulfilling the arms pledge. On the same day de Nerva, after his return from Cairo, confirmed to the press in Bonn the ending of the West German arms deliveries to Israel but added that the Federal Government had promised not to recognise Israel in the near future. This latter statement caused consternation and anger in Bonn where the impression was given that it had been made without the government's authority. It is believed, however, that such an assurance was indeed given by the Foreign Minister, Gerhard Schröder, who was known to believe that the non-recognition of Israel should continue if a breach of the Hallstein Doctrine by the Arab states was to be avoided. Erhard had at that moment made no decision on diplomatic relations with Israel and, apart from impeding the Chancellor's freedom of action, de Nerva's statement was bound to add further fuel to the anger of the Israelis. It was denied by the Secretary of State of the Foreign Office, Carl Carstens, during a Bundestag debate on the Middle East nearly a week later, and not until after Eshkol had delivered a strong attack on the Bonn government for stopping the arms deliveries. In Israel the arms agreement, whether secret or not, was regarded as binding.

West German anger with de Nerva's performance – he was not received by either the Chancellor or the Foreign Minister – may have been due to tactical considerations rather than to the substance of his statements: the Spanish mediator had used undue haste in completing his mission and in announcing his success. The Federal Government, which had long been divided on the issue of the arms agreement, had not yet agreed on the final formula and was still negotiating with Israel, hoping to persuade the Israeli government to accept a financial arrangement in place of the arms deliveries still outstanding. Neither had a final attitude towards the recognition of Israel emerged. Here much could depend on the development of relations with Egypt and whether the Federal Government would succeed in persuading the Egyptian President to call off the Ulbricht visit or at least to tone down its importance. The rapid revelation of the arms embargo before the negotiations with all the parties had been concluded and agreement had been reached within the government had undoubtedly caused disarray. The Federal Republic was now in trouble with both the Arabs and Israel and it was not clear what West German policy was achieving as it became obvious that no changes would be made to the Ulbricht visit. De Nerva's only reference to the visit on his return from Cairo had been that the UAR would 'make several gestures of goodwill, which will become known in time'.[34] Only the threat that Egypt would recognise the German Democratic Republic seemed for the present to have passed, at least Ali Sabry had made no mention of it when he announced the West German arms stoppage in the Egyptian parliament. But the price paid by Bonn for its policies in diplomatic terms was certainly high.

In his statement to the Knesset on 15 February Eshkol made it clear that he expected West Germany to fulfil its obligations in the spirit and to the letter and that Israel rejected outright any form of compensation. Germany's primary moral duty was to make every possible contribution to the strengthening of Israel. He wondered whether the West German government had not considered that its surrender to Cairo would start a chain reaction of extortionate threats from other quarters and that that would increase international tension. Eshkol rejected the West German claim that Israel was one of the world's 'areas of tension'. There was no basis for this definition in the case of Israel because the situation there resulted from the Arab policies of threats and aggression, which justified Israel's right to assistance. Eshkol also contrasted the haste with which the Federal Government had acted to halt arms to areas of tension with its sluggishness over solving the problem of the West German scientists in Egypt. Nevertheless, he also said that he did not regret any aspect of Israel's past relations with the Federal Republic. Israel wanted to maintain relations with West Germany and not burn all the bridges.. Most of the Knesset, Eshkol continued, hoped Israel's relations with West Germany would continue and improve. The Knesset debate was followed by a resolution, passed by 54 votes to 26, expressing shock and indignation at the Federal Government embargo on the security aid promised to Israel. This intention, if carried out, would amount to surrender to the Egyptian ruler's policy of hostility to Israel and to an encouragement of its implementation. In view of this Israel would have to reconsider the significance of Israeli–West German relations. The Knesset approved the government's demand that Bonn should fulfil its obligations in the spirit and to the letter, and its refusal to accept any financial compensation as a substitute for the cancellation of the security aid.[35]

The Israeli press had been apprehensive from the start of the crisis, and indignant once the end of the arms deliveries had been announced by the Erhard government. In an early reaction to the West German–Egyptian conflict over the Ulbricht invitation the *Jerusalem Post* wrote 'From the moment the Federal Republic adopted the Hallstein Doctrine ... it laid itself open to the sort of blackmail now being attempted by Nasser', and in a swipe at the Erhard government 'Adenauer's Germany was not easily blackmailed or pressed into actions it did not want to take'.[36] The left-wing Labour *Al Hamishmar* took a similar line, but added 'Bonn continues to delude itself that it will succeed in strengthening its position in Egypt and the Arab countries at the expense of its debt towards Israel'.[37] In a leading article the *Jerusalem Post* commented 'The surrender of the Erhard Government to Nasser's ultimatum has been so abject as to seem incredible at first sight ...'. The paper criticised the Social Democrats, who in the past had professed friendship with Israel, for not protesting. From a military point of view, it continued, West German military aid was not crucial as it could be acquired by other means 'But the blow to prospects of ultimately putting relations between Germany and us on a firm footing will not be so easily overcome'. To do so had been the dream of Adenauer and Ben-Gurion, who looked to the future. In a column of foreign relations comments the paper wrote under the heading 'Bonn's Moral Collapse': 'Nasser's gamble has paid off. By courting Pankow[38] he has cowed Bonn ...'. Israel, of course, was the immediate loser. But the implications went further for by its policy the Erhard government seemed not only to have shattered the prospects of normalising relations between West Germany and Israel, but exposed itself to wholesale international extortion.[39] *Davar* (Trade Union supporter of the government) commented

> The Bonn Government has no alternative but to decide whether it will meet its obligations, which arise not only from a formal agreement but also from its moral duty, or whether it is prepared, even unintentionally, to side with Egypt against Israel.[40]

Following Eshkol's Knesset speech most Israeli papers agreed with their Prime Minister's contention that the Federal Republic should be urged to fulfil its obligation but some were cautious, as Eshkol had been, about future West German–Israeli relations. The *Jerusalem Post* commented

> Mr Eshkol has not excommunicated the Germans and he has not pardoned their breach of faith, nor has he given much indication of how our limited relations with Germany are to develop in the future .... One of the main elements to be considered will be the fact that Germany is already again one of the most powerful states in Western Europe and that it will be sufficiently difficult for us to achieve the kind of association we wish for with European organisations without making the inconvenient condition that we must be permitted at all times to boycott Germany.[41]

Chancellor Erhard, in a long statement to the Bundestag on 17 February 1965, expressed disappointment with the Israeli reaction. He reiterated that for Germany there remained the debt to the Jewish people which the Third Reich had imposed on the German nation, but he had hoped for greater understanding by Israel for the situation in which the Germans found themselves as a result of the division of their country. In a reference to the offer to compensate Israel financially for ending the arms deliveries he said: 'We did not unilaterally break an undertaking but simply proposed that we might discharge the remainder of our obligation in such a way that Israel would suffer no material disadvantages.' Erhard had some harsh words for the actions of the Egyptian President:

> The UAR appears to want to ignore the most vital questions of our existence. We have always proved by our deeds that we are serious about the preservation of our friendship with the Arabs. We therefore have a right to ask what proof there is of Egyptian friendship. Those who treat Ulbricht as the head of a sovereign state make a deal with those who split the German nation.

President Nasser, Erhard continued, could not fail to realise that West

Germany's relations with Egypt would be most seriously affected by Ulbricht's visit to Cairo. He then declared that Bonn would end its economic aid to Egypt and would reserve the right to take political steps also.[42]

With this statement West Germany's Middle Eastern crisis was moving to a climax. After Erhard's threat to cut off development aid and possibly diplomatic relations a showdown with Egypt seemed inevitable. At the same time the Federal Government was now in conflict with Israel, which resented the assertion that it was situated in an area of conflict and therefore not eligible to receive military aid, and insisted on the fulfilment of the secret arms agreement, refusing to accept the West German offer to convert it into financial aid. Nasser, on the other hand, remained defiant over his invitation to Ulbricht. At a rally at Aswan he declared that he would not cancel the invitation, whether the West German government liked it or not. Egypt would sever relations with the Federal Republic if the latter continued to supply arms to Israel. He objected, he said again, to the arms supplies because they were a gift and the deal had been concluded secretly. He rejected West German threats to cut off economic aid to Egypt if Ulbricht visited Cairo.[43] Nor apparently did the Federal Government receive much solace from its friends. In his Bundestag speech Erhard had expressed the hope that he could count on the solidarity and support of West Germany's allies over the German problem and its current extension to the Middle East, but the American government, while going so far as to admit that it had been consulted and had approved the transfer of American-made tanks by the Federal Republic to Israel,[44] otherwise remained silent. At this juncture there seemed no way out of the stalemate and the Federal Government was very much alone.

There was concern about the West German abrogation of the arms agreement with Israel and severe, sometimes scathing criticism in some western newspapers of Bonn's handling of the crisis. The London *Observer* stated that the Federal Republic should have considered the implications for the Hallstein Doctrine before proposing to become involved in the Middle Eastern arms race, but once involved it should have stood firm over the arms deliveries to Israel. The result of its wobble in the Middle East was to increase tensions there and promote a sense of betrayal among the Israelis.[45] Nasser's prestige would be enhanced among the Arab states, toughening their intransigence towards Israel.[46] The *New York Times* in a leading article, criticised the West Germans' 'almost incredible mess up' of the linked issues of arms supplies to Israel and their relations with Egypt. As a result Nasser had been greatly strengthened and Middle Eastern tensions were rising. West German–Israeli relations had sunk to a bitter low while East Germany was achieving a triumph. The Federal Republic should have continued the arms agreement once made – even if mistakenly – to sustain the Middle Eastern arms balance. On West Germany's threat to cut its economic aid to Egypt, the paper commented that Nasser had proved in the past that he was allergic to dictation based on aid. Meanwhile Soviet arms supplies to the Arabs were continuing.[47] The Amsterdam *Volkskrant* talked of 'lack of leadership, self-assurance, character, moral fibre, style and posture', the Paris *Combat* of the roar of the rubber lion Erhard and *Aurore* wondered whether the Germans still had a Chancellor.[48] The *Daily Telegraph* believed that Bonn's threat to cut off economic aid to Egypt if Ulbricht was received there showed that 'there are at least some limits to the German government's ineptitude in this matter'. Despite Nasser's blackmail, however, the Federal Republic should complete its arms deliveries to Israel as the Israeli government was demanding. 'As has already been proved, submission to Nasser's blackmail is as unprofitable as it is dishonourable.'[49]

In the Federal Republic too, criticism was severe. The SPD Press Service called the situation a 'diplomatic Stalingrad'.[50] The liberal weekly *Die Zeit* asked why Erhard did not immediately stop the arms consignments before pressure was applied from Cairo, recognise Israel and threaten economic sanctions; doing nothing had helped nobody. 'We were caught between the pipers of Cairo and Jerusalem. Why did not Adenauer just pay for the arms instead of delivering them?'[51] The left-wing *Abendpost* wrote that the 'so-called statecraft' of Bonn's policy had sunk to absolute zero. The Federal Republic was despised in Israel, humiliated by President Nasser, mocked by the rest of the world.[52] The *Handelsblatt* declared 'Now apparently we have as good as lost the Middle Eastern game on both sides',[53] and the independent *Stuttgarter Zeitung*:

> German dignity and German reputation have sunk in the waters of the Nile .... While clowning, calling itself German Middle Eastern policy, is continued, Cairo and East Berlin publish the timetable for Ulbricht's visit to the Egyptian capital.[54]

Die Welt believed that the West German threat to cancel aid to Nasser if Ulbricht's visit was not called off had helped to improve the Federal Republic's chipped image but feared that it had come too late. Eshkol's Knesset speech that Israel would not accept money instead of weapons deliveries showed how much ground had been lost. It was imperative that a crisis with Israel should be avoided.[55] The *Frankfurter Allgemeine Zeitung* was one of the few that agreed with the government, saying that by ending the arms deliveries to Israel and threatening to withdraw aid to

Egypt if it did not cancel the Ulbricht invitation the Federal Government had corrected its earlier mistakes and won itself freedom of movement,[56] but it also criticised the government in a leader for its 'vacillation and annoying unclarity' as a result of which it would spoil its relations with both sides.[57] If one could talk of compromise the Federal Republic had obtained very little and was paying a lot and might pay more. It had given in to political blackmail and it would be difficult to resist further pressure.[58] The weekly news magazine *Der Spiegel*, sub-titling its article 'The Federal Republic has turned Jews and Arabs into enemies: the world laughs at Bonn', summed it up in this way:

> The Bonners are helplessly exposed, as prisoners of their own Hallstein Doctrine, to a see-saw policy of this calibre. Whenever a state of the Third World threatens to recognise Pankow, German development aiders travel around as motorised Santa Clauses, pouring out their sacks of gifts. Nowhere was the Hallstein Doctrine so expensive as on the Nile.

Over the question of Israel Nasser had not recognised a see-saw policy. However much he would like to play one German state off against the other to obtain foreign currency, he was firm over Israel. On the other hand the Germans could not treat the Jews like any other people. Hence the Hallstein Doctrine had led Bonn's foreign policy into a blind alley. Nasser had answered Bonn's demand not to recognise the German Democratic Republic with a counter doctrine: if the Federal Republic established diplomatic relations with Israel, he would recognise the GDR.[59]

The Ulbricht visit to Cairo took place as planned from 24 February to 2 March. There was little sign that it had been toned down, indeed there was every indication that the First Secretary was accorded a full-dress welcome by the Egyptian President. As had been expected, the GDR leader brought with him an offer of a $100 million loan and the promise of a trade agreement as well as fraternal greetings and a return invitation to President Nasser to visit East Berlin. What the Egyptian President, for his part, had to offer to his East German guest may have been less than that which the visitor had hoped. Ulbricht did not obtain full recognition of the GDR, nor a promise that this would be extended to his country in the near future. Though he had many meetings with Nasser, the latter, while extolling the good relations existing between the two countries and the virtues of socialism, insisted strongly in his speeches on the non-alignment of Egypt and of the Arab world in general. By stressing his neutrality in the power struggle between the communist and the western world he was implying that he could give only qualified support to the GDR in its own continuing cold war with the Federal Republic and its fight for international recognition. He reinforced this, saying at a state banquet held in honour of his guest that he was 'still exerting our maximum and most sincere efforts so that matters do not deteriorate any further' in Cairo–Bonn relations. He added that Egypt had always been careful to maintain good relations with Federal Germany, but that the Israel arms deal had been a stab in the back.[60] The official joint communiqué, issued just before the end of the visit, said little of importance. Ironically, it asserted the vitality of the 'traditional German–Arab friendship', a formula so often repeated by both the Egyptian and West German governments, most recently in Chancellor Erhard's statement to the Bundestag. But whereas Ulbricht gave support in the communiqué for Egypt's policy of non-alignment and Nasser's efforts on behalf of Arab and African unity and the rights of the Palestinians, the President, while expressing sympathy for the German people, contented himself with saying that the question of German unity was a matter for the German people, thus reiterating a principle adopted at a conference of non-aligned states that divided countries should strive to reunify their territories by peaceful means 'without interference or pressure from outside'.[61] This could not have brought much solace to the head of government of the GDR.

That Ulbricht's success in Cairo was limited gave little comfort to Bonn, despite the fact that the threat by Egypt and some other Arab states to recognise the GDR seemed for the moment to have passed. The Federal Government, venturing for the first time beyond Europe into world affairs, had suffered a serious diplomatic humiliation. It now had to make hard decisions: firstly to try to mollify the Israelis who, at the end of a line of criticisms levied against the Federal Republic over diplomatic relations and other matters, felt deeply affronted by what they considered to be a breach of an agreement and a surrender to blackmail; and secondly to save what was left of West German influence in the Arab world and its modest, but by no means unimportant, role as the ambassador of the West which, by its popularity, had contributed to restricting Soviet influence in the Middle East. Of these two problems, the latter was the more difficult to solve.

Regarding the Arabs Bonn had achieved nothing but a retreat. The future of the Hallstein Doctrine, together with the entire reunification policy as understood by the Federal Government, was now more than ever in doubt, since the threat to recognise the German Democratic Republic could be applied again in an attempt to force a change in the Federal Republic's foreign policy. The doctrine had been upheld mainly by the bribe of development aid, which had made it tempting for many Third World countries to obey Bonn's injunction not to have dealings

110

with the East Berlin government. The Federal Republic had been pre-pared to pay heavily for this in the hope of preserving its chances of ultimate reunification through strength, a policy seen increasingly both inside and outside West Germany as a mirage. President Nasser had rejected the bribe and declared his intention to dispense with West German offers of even more help. With this the Federal Government's use of economic aid as an instrument of foreign policy, widely applied at the time, seemed close to collapsing.

Early in March, that is shortly after Ulbricht's return to East Berlin, Erhard was still toying with the idea of the Federal Republic breaking off diplomatic relations with Cairo. This had been foreshadowed in Erhard's speech to the Bundestag on 17 February when, after announcing that the Ulbricht visit would mean the end of West German economic aid, he added that the Federal Government reserved the right to take 'political steps' also. The Chancellor's anger was further fuelled by Nasser's announcement that Egypt would in due course open a consulate-general in East Berlin. There was nothing revolutionary in this, since an East German consulate-general already existed in Cairo. Erhard was now strongly opposed by many of his cabinet colleagues, especially the Foreign Minister, Gerhard Schröder, as a break in diplomatic relations with Egypt would be tantamount to the abandonment of the West German position in the most important state of the Arab Middle East. To under-stand that the vacancy left by the departure of the West German ambas-sador would be quickly filled by an East German needed little imagination. Schröder received the support of the ministers belonging to the FDP, the smaller of the two coalition parties in the government, while Erhard received strong backing from the CDU, including the still influential CSU leader and former Defence Minister, F.J. Strauss. There was now an open split in the cabinet and the risk of a government crisis, which did not improve the chances of sound decision making, so important at that moment. What caused Erhard to relent was, apart from the opposition in his cabinet, the advice of the American ambassador, McGhee, who made it clear that Washington was not interested in a new crisis area in view of the American preoccupation with Vietnam. The British and French ambassadors, though more cautious, were also reported to have advised against a West German breach with Cairo.[62] Bonn's actions against Egypt were therefore confined to the ending of economic aid, though it was stressed that any upgrading of the GDR would lead to further measures by the Federal Government. The matter did not, however, end there.

The Erhard government, having stated its terms to the UAR, now had to tackle its crisis with Israel. The problem here was equally complex but also more capable of a solution and the outcome was therefore more predictable. That the Israeli government and parliament had rejected the West German offer to convert the outstanding arms deliveries into financial aid cannot really surprise, not only because of possible Israeli difficulties with finding a substitute supplier of weapons, but even more so because of the continuing Israeli attitude that they can expect more from the Germans than just political or business deals that are advantageous to both sides: that the Germans must take responsibility for the survival and well-being of the Jewish State. The Erhard government's task was complicated at this stage by the other complaints the Israelis were levying against it: over the non-existent diplomatic relations, the scientists in Egypt and the still unresolved problem of the Statute of Limitations in relation to war criminals. Only one of these, the establishment of diplomatic relations, offered the possibility of immediate action which might soothe the anger of the Israelis. But the circumstances which had prevented normalisation, namely the fear that the Arab states might react by recognising the German Democratic Republic, had not really changed. In some respects the anger on both the West German and Egyptian side had increased the threat, while on the other hand there were straws in the wind indicating that Nasser did not wish to take the final step and cut himself off even further from the West: his moderate stance towards Gerstenmaier during the latter's Cairo visit the previous November and his reception of Ulbricht as described in the final communiqué issued at the end of the visit. Erhard was now in favour of the immediate establishment of full diplomatic relations, but again met with opposition from some of his cabinet colleagues. This time it was his view that prevailed.

Erhard's special envoy, Dr Kurt Birrenbach, who was sent to Israel to negotiate on both diplomatic relations and the conversion of the arms agreement, tells of the events in some detail in a commemorative article, published in honour of Erhard some years later.[63] Birrenbach's qualification for the mission, it should be said, was not a special commitment on his part to Israel, but rather his wide international business experience and, above all, his contacts with important political and economic personalities in the United States.[64] It was from there that he was recalled urgently by Erhard to undertake his mission, after having discussed the implications of the Federal Republic's current crisis with American politicians and members of the Jewish Communities. It was represented to him that the US considered it important for the West that the Federal Republic should not be excluded from the Middle Eastern danger zone since it was a moderating factor.[65] At the same time Jewish community leaders cannot have failed to impress upon him their dismay at the West German abrogation of the arms agreement.

Three meetings with Erhard were held, attended by Birrenbach, Foreign Minister Schröder, Defence Minister von Hassell and other lesser government officials. At the first, three possibilities were envisaged: the immediate establishment of diplomatic relations with Israel; the opening of a consulate-general, to be upgraded to an embassy as soon as possible; and the continuation of the status quo. There was little support for the third proposition, but a majority, including Schröder, was still opposed to the first idea, strongly advocated by Erhard and supported by the British, American and French ambassadors in Bonn, that the time was ripe for full diplomatic relations. At the first two of these three meetings no decision was reached because of disagreements, but even at the third, which was attended by Rainer Barzel, the chairman of the CDU parliamentary group, immediately upon his return from the US, Erhard's views did not prevail. Birrenbach was therefore sent away with instructions to offer the Israelis a consulate-general, despite repeated statements by Israeli politicians, including Eshkol, that nothing short of full diplomatic relations was acceptable. The only mitigating factor was an assurance that the consulate would ultimately be converted into an embassy; the hope expressed at the Bonn meeting was that in the meantime steps could be taken to improve the relations with the Arab states.[66]

It was undoubtedly the advice of the three ambassadors and the message Barzel brought from New York, especially the latter, that decided Erhard to take the plunge. Barzel had met the Israeli ambassador to the US and the United Nations, Harman, who warned him that a diplomatic victory by Nasser over the Federal Republic would endanger Israel and had also gained the impression from his American contacts that the crisis in the relations between the Federal Republic and Israel would have unforeseeable consequences.[67] The West German action of stopping the arms flow to Israel, seen as submission to blackmail, was severely criticised in the United States. It was regarded by some as a setback to American policy in the Middle East which would benefit the Soviet Bloc, and by the Jewish communities as a serious let-down of Israel. Some Jewish organisations instituted a boycott of West German goods.[68] The general atmosphere and the implications for West German–Israeli relations could not long have remained hidden from the Bonn government. Barzel who, like Erhard and Adenauer, was known to have been conscious of the strong moral involvement of the Germans in Israeli affairs, is believed to have thrown his personal views into the scales. On 7 March, a day after Birrenbach had left on his confidential mission armed with instructions to offer only a consulate-general, the Federal Government issued a long statement on the Middle Eastern situation, which contained a paragraph saying that the Federal Republic was now 'aiming at the establishment of diplomatic relations with Israel'.[69] Erhard had made the decision against the advice of his Foreign Minister and several other cabinet colleagues, using his prerogative as Chancellor, given him by the West German constitution, to determine the general directions of policy. His emissary, Birrenbach, was interrupting his flight in Zurich at the time; he was informed of his government's change of heart only on his arrival at Tel Aviv, ironically by his future negotiating partners in the Israeli government.

In offering Israel full diplomatic relations, Erhard had certainly taken a bold step. What his feelings may have been when he made a decision that had been postponed for nine years has not been revealed. That he could have despaired of the Hallstein Doctrine and the Federal Republic's aspirations regarding reunification is belied by his anger about the Ulbricht invitation, which showed again in the government declaration of 7 March: the Ulbricht visit was still regarded as an affront to the Doctrine. A more plausible explanation would be that the Egyptian President had shown by his actions that he would pursue his policies whatever counter-measures the Federal Government might take, and that West German influence would not be strong enough to divert him from the course of recognising the GDR if he wished to do so. Finally, the Chancellor must have known that if he wanted to save West German–Israeli relations, and this was certainly his intention, the offer of a consulate at this moment would have been counter-productive. Erhard, if he still believed in the Hallstein Doctrine, certainly took a risk. Nasser's attitude to the Federal Republic's normalisation of relations with Israel, in the still relatively calm atmosphere of the Gerstenmaier visit to Egypt of November 1964, could not easily be maintained after the fierce exchanges between Bonn and Cairo following the Ulbricht visit and the ending of West German economic aid to the United Arab Republic. For a man who aimed at leadership of the Arab world, who by his style and actions had made himself the champion of the Arab masses, this was not the moment openly to soft-pedal on a policy he had proclaimed for many years. His reaction to the announcement that Bonn intended to establish diplomatic relations with Israel was now that if this were confirmed his and other Arab governments would recognise the GDR. Like the rest of the world he may have been taken by surprise by Erhard's decision, hence the hesitation. But once the confirmation did reach him he hastily convened a conference of Arab foreign ministers in Cairo at which he and several other Arab governments proposed recognition of the GDR. That the meeting ended without it is due to the strong opposition of a number of Arab states to such a step. The communiqué issued at the end of the conference merely stated the decision to recall the Arab ambassadors from Bonn and to break off diplomatic relations with the Federal Republic. Even then there was no unanimity: three Arab states, Libya, Morocco and Tunisia dissociated themselves from the Arab majority and refused to comply.

The Birrenbach discussions in Jerusalem were secret but many details have transpired since. Once the nature of the West German offer of diplomatic relations had been clarified – and the wording of the German statement that the Federal Republic was 'aiming at' diplomatic relations with Israel raised some doubts in the minds of the Israeli negotiators at first – a number of other points became clear: the West German government would not go back on its decision not to send arms to areas of tension, which would include Israel; it was prepared to be generous in its offer of financial compensation; the other outstanding problems between the two countries, the German scientists in Egypt and the Statute of Limitations would also be discussed and were capable of solution; and the Israeli government, whatever might be the fate of the talks on the arms agreement, was not in a position to reject Erhard's offer of diplomatic relations. Shinnar lays stress on the decision of Israeli Prime Minister Eshkol to separate the issue of diplomatic relations from the remaining issues, that is, to accept Erhard's offer of normalisation first and then to continue to negotiate about the more complex problem areas.[70] It would indeed have been difficult, in view of the resentment felt in Israel over the arms embargo, to have sold to the Israeli parliament and public a whole package which would have included the conversion of the arms agreement into a financial arrangement. Israel had for many years urged the West Germans to take the initiative over normalisation of relations. The Israeli government had taken this line for reasons of national interest, that is, in recognition of the Federal Republic's growing economic power and political influence in Europe, the area with which Israel for cultural, political and economic reasons had the closest affinities. As members of the Israeli government had repeatedly stressed, the Federal Republic's good relations with France, at that time Israel's closest, some would say only, ally were of great importance. There were now, in addition, tactical reasons for a rapid solution of the diplomatic relations issue: the absence of full relations had become an Arab propaganda point increasingly damaging to Israel[71] and the government in Jerusalem, conscious of past West German prevarications, thought it prudent to accept the Federal Government's offer quickly before it weakened again in the face of Arab pressure. Acceptance could strengthen the hands of the Federal Government against the rising tide of Arab abuse. Eshkol therefore declared his readiness to press for an acceptance of diplomatic relations with Bonn in the Knesset and was supported in this by his cabinet. The acceptance was conveyed by the Israeli government to Bonn on 16 March and endorsed by the Knesset the same day by 66 votes to 29, with 10 abstentions.

115

Over the question of the arms agreement progress was far less easy, as the Israeli side showed no willingness to compromise at first. The West German statement that Israel was an 'area of tension' continued to be disputed. The Israelis have over the years of conflict with their Arab neighbours considered their country as a stabilising factor in the Middle East. Only the Israel Defence Forces had, in their view, prevented or at least limited major wars in the area. A more immediate difficulty for the Israeli government was that Eshkol had made a public statement rejecting the West German proposal to stop the arms deliveries and pay compensation instead; this statement had been endorsed in the Knesset by a majority vote. On the German side there was of course the over-riding fear that the Arabs would carry out their threat to recognise the GDR, but more long-term, the reluctance to becoming involved in international disputes. There was still an inclination, dating from the end of the Second World War, to adopt a low profile in international affairs by contrast with some other west European states. The Federal Republic's modest incursion into the politics of the Middle East, the most dangerous of conflict areas, had been an exception and had now led the country into a serious predicament. All shades of political opinion in the Federal Republic had disliked and often severely criticised the arms agreement with Israel once it had become public knowledge. Erhard could count on massive support at home for its abrogation. There was no chance therefore of the Federal Government changing its resolve not to send arms to areas of tension in the future. As a result the talks in Jerusalem became deadlocked, with the danger that they might break down irrevocably.

The solution was facilitated by the fact that the arms that were no longer to be delivered from West Germany could be obtained elsewhere. The American government had been in favour of Israel receiving weapons so that an arms balance could be preserved in the Middle East as long as the Arab states were getting large quantities of military supplies from the Soviet Bloc. But Washington did not want to be seen arming Israel for fear of losing Arab goodwill. In reality most of the weapons supplied by the West Germans were of American manufacture. Somehow the Americans now had to step into the breach. At the height of the crisis, in late February and early March, the American diplomat Averell Harriman was in Israel as a special envoy of President Johnson, trying to soothe Israeli anger with the Federal Republic over the ending of the arms deliveries and negotiating, it was believed, a replacement of these by the United States. The negotiations were complicated by another Middle Eastern dispute in which Israel was deeply involved and which arose due to a plan by Jordan, Syria and the Lebanon to divert the headwaters of the River Jordan, much of which flows through Israel and is an important source of water, especially for Israeli agriculture. The Israeli government, which regarded the Jordan waters as vital for the existence of the country and feared that diversion by the Arab states was an attempt to destroy Israel, threatened military action against the Arab states involved if the plan were carried out. There was some international disagreement about the seriousness of the situation Israel would face if the Jordan waters were diverted and Washington, unwilling to face another Middle Eastern war while its hands were tied in South-East Asia, tried to play down the threat that was hanging over Israel. Although no statement was issued about the content of the negotiations at the end of the Harriman visit, it is believed that the American drove a hard bargain: any American arms deliveries to Israel would be dependent on Israeli restraint over the Jordan waters. The Israeli government did not publicly withdraw its threat to attack if the waters were diverted,[72] but the plan was soon abandoned by the Arab states concerned. American arms, on the other hand, were delivered to Israel.

Agreement in principle was finally reached as a result of a concession by Israel, when it became clear that American arms deliveries could be obtained. The Israeli government was able to accept the conversion of the original arms agreement with Bonn into a financial arrangement because the weapons would continue to arrive and the Federal Republic would in practice pay for them. In this way Israel would suffer no disadvantage, since its security requirements would continue to be fulfilled and no new financial burden would be placed on the Israeli exchequer. But the talks, which lasted for about six weeks and were interrupted twice to enable Birrenbach to travel to Bonn for consultations and further instructions, ranged over a number of other subjects including the establishment of diplomatic relations, the future of the economic aid agreement entered into between Adenauer and Ben-Gurion in 1960, the West German scientists working in Egypt, the Statute of Limitations and its bearing on German war criminals and the prospects of an association agreement between Israel and the EEC, a matter of immediate interest to the Israeli economy and on which negotiations were proceeding at the time. None of these issues was, however, settled by Birrenbach, who had not been empowered by his government to conclude final arrangements. Instead it was agreed to continue negotiations on most of the issues through normal diplomatic channels once these had been established in the course of the summer of 1965.[73] Nevertheless agreement was now assured. On 12 May letters were exchanged between Chancellor Erhard and Prime Minister Eshkol, setting out the main points on which agreement in principle had been reached. Concerning the scientists in Egypt, Erhard in his letter assured the Israeli Prime Minister that a large number of these had now returned to West Germany and that a further number, especially those engaged in rocket construction, also intended to return soon. The West German authorities would use all the legal powers at their disposal to take action against individuals who recruited West German nationals without permission for scientific or technical work in the military field outside West Germany.[74] Concerning the Statute of Limitations there had meanwhile been a decision by the Bundestag to extend the period during which war criminals could be brought to justice by four years up to 1969.[75] Finally, on the question of association with the EEC Birrenbach was able to assure the Israeli government that Bonn would strongly support Israel in its bid for an agreement.[76]

In Israel the West German offer was at first given only a qualified welcome. The main reasons for this were the circumstances in which the Erhard decision had been taken. There were fears that since the Federal Government had been manoeuvred through its own policies into a position where it was forced to offer diplomatic relations the offer might not be genuine and the new relationship might be unsatisfactory. The *Jerusalem Post* wrote in a leader 'We really do not wish for diplomatic relations with any state as the by-product of its dispute with a third power'. There were, the paper added, immense historical and personal obstacles to relations between Jews and Germans only 20 years after the war.

> The fact that Germany for reasons of its own political convenience declined to initiate formal relations with us became an added insult and an added difficulty to the development of practical contacts where they were desirable. . . . if there are to be any relations with the Germans they will have to realise the special character of the connection . . . . Either they believe that this is right and essential for the direction which the future Germany is to take and therefore no less important than other political considerations, or they view it as expendable, in which case it is not worth attempting.[77]

But the trade union paper *Davar*, which supported the government, wrote 'Even those who approve the setting up of official relations with Germany underrate the fundamental importance of Bonn's initiative'. The decisive consideration from Israel's point of view should be the interests of the nation – of its international status and security.[78] It added a few days later that Bonn's proposal on diplomatic relations 'is an important achievement for Israel in its trial of political strength with Arab countries. We have no right to forfeit this achievement and bring about a situation in which it could become an asset to the Arabs.'[79] *Haboker* (Liberal) said that the question of relations called for frank clarification, especially after Bonn had broken not a few agreements with Israel.[80] The independent *Ha'aretz* agreed with the Prime Minister that there was no point in linking the question of diplomatic relations with the arms issue or demands for economic aid or other conditions.[81] *Hatsofe* (National Religious Parties) urged the government to take 'categorical decisions' on the West German proposal to establish diplomatic relations without waste of time and energy on superfluous discussions. 'The establishment of relations with West Germany, which enjoys such an important economic and political position, holds prospects of increasing Israel's deterrent force on the one hand and driving a wedge into Arab unity on the other.'[82] The right-wing opposition paper *Herut* was predictably scathing: even if Erhard, due to Nasser's political game, felt compelled to propose diplomatic relations with Israel,

> we should divert our attention from the ominous consequences of official relations with Germany to the very soul of the nation, especially since everyone knows and understands that the motive for the Chancellor's decision was not pangs of conscience for the destruction of six million Jews but the need to settle accounts with a faithful friend, the Egyptian dictator .... Does anyone believe the resentment between Bonn and Cairo will last for ever?[83]

The Prime Minister's arguments in favour of acceptance and the Knesset vote, however, helped to dispel many doubts. A leading article in the *Jerusalem Post*, noting Eshkol's and the Knesset's acceptance of the offer of diplomatic relations with the Federal Republic, stated:

> If it is indeed as has been said, a victory of reason over emotion, it is not an easy victory. From every practical and contemporary point of view official relations with Germany are necessary for the strengthening of the State, which cannot persuade the rest of the world to leave Germany in quarantine for another half century until the last Nazis of Hitler's day have died out. To sacrifice even a part of Israel's security and stability in order to be able to reject Bonn's proposal indicates, more than anything else, a false estimate of the nature of diplomatic relations ....

There could be no doubt that Germany as a whole, and especially the younger generation, had turned away from its two-generation dream of world domination and military glory and had joined the rest of Europe for more tangible and more easily achieved goals. To attempt to isolate the Germany of today, including those who strove to rehabilitate it morally and intellectually would not contribute to the eradication of Nazi mental processes but might appear to justify them.[84]

The trend reflected the by now familiar struggle between emotions and raison d'état which had always characterised the reactions of the Israeli public and media to any decision over an aspect of German–Israeli relations but in which reason ultimately won the day. The struggle was evident in the Knesset debate where the opposition parties on both the far right and far left had dwelt strongly on the Holocaust in persuading members of parliament and of the government that there could be no forgiveness or reconciliation and therefore no relations with any Germans. But the majority, many of whom were undoubtedly affected by the emotions of the hour and for whom the decision was therefore a difficult one, ultimately voted in favour of diplomatic relations, persuaded that their establishment served the interests of the State. As the dust settled after the public debate, however, many became aware of the practical implications: a German ambassador, and how he could be protected from hostile elements, the German flag to be hoisted in Israel for the first time, the German national anthem to be played by an Israeli military band to welcome the new ambassador; but all of it together would be, as *Davar* put it, 'the price that emotions must pay to political reason, since the nation has chosen to live a life of political sovereignty'.[85]

Meanwhile in the Federal Republic pressure of public opinion for diplomatic relations with Israel had been mounting during the crisis, despite discontent with and criticims of the secret arms agreement. An open letter signed by 435 university teachers in the Federal Republic was sent to Schröder, Gerstenmaier and members of the Bundestag's Foreign Affairs Committee, pointing out that the unwillingness to establish diplomatic relations with Israel and the work of German scientists and technicians in Egypt had seriously impaired relations between the Federal Republic and Israel and between the German and Jewish peoples. This letter, first sent in November 1964 and followed by a reminder in February 1965, was inspired by a well-known Lutheran theologian, Helmut Gollwitzer; it may have indirectly influenced Erhard's decision of 7 March.[86] A petition signed by workers and employees was sent by the West German Trade Union Federation, appealing for diplomatic relations with Israel and warning the government not to be deflected from the 'righteous course' by threats of other states or by financial considerations. Once Erhard's decision to establish diplomatic relations with Israel was announced, it was welcomed by the majority of political and public opinion. The main reservations came from the Foreign Minister, Gerhard Schröder, and the Free Democrats in the cabinet and the Bundestag because of fear for future West German relations with the Arab states. The CDU parliamentary group, however, declared its full support for its leader and a CDU spokesman, hearing the Israeli response, called Israel's acceptance 'logical and gratifying'.[87]

But the offer of diplomatic relations with Israel was largely overshadowed by the many other aspects of the crisis: the Federal Government's see-saw policy in the Middle East, the lack of decisiveness and the lengthy deliberations since the crisis began in October 1964, the rift in the cabinet and above all the threat by some Arab states to recognise the GDR, the West German position in the Middle East and the future of the Hallstein Doctrine. That was a formidable list of woes. In it the proposed normalisation of relations with Israel was near the end of the queue and references to it in the press, almost all favourable, were tinged with concern that the Israelis might reject it or try to trade it against unacceptable demands such as the continuation of the arms deliveries. The right-wing *Die Welt*, known for its persistently pro-Israel attitude, while agreeing in general terms with Erhard's package of decisions of 7 March, deplored the arms stoppage, saying that 'the abrupt sacrifice of the undertaking to Israel, made on the altar of German–Arab relations, has seriously shaken the German position'.[88] Concerning the new relationship with Israel, it added that in the epoch which had now begun in German Middle Eastern policy the German attitude would be freed, if Jerusalem agreed, from the twilight in which it had always been placed. After the Israeli acceptance it wrote:

> The impending establishment of diplomatic relations with Israel is the only positive element in the crisis of German Near East policy. The readjustment of our relationship to Israel is politically long overdue. As the last of the larger Western states the Federal Republic is now establishing the only appropriate relations with Israel.[89]

The left-liberal *Frankfurter Rundschau* similarly welcomed the fact that

> the crisis of our Near-East policy has created one positive breakthrough; at last our relationship with Israel moves out of the twilight and takes on clearer contours. It is to be hoped that the dialogue in Jerusalem will not fail over emotional encumbrances.[90]

The *Süddeutsche Zeitung* welcomed the decision to establish diplomatic relations with Israel but regretted the opportunities that had been missed: 'For years this paper asked and angrily complained about why this decision was not taken earlier. It would have been swallowed more easily by the Arabs.'[91] But the paper also criticised the decision as 'a by-product of the dispute with Nasser rather than the expression of a positive attitude'.[92] In the case of the *Frankfurter Allgemeine Zeitung* relief that the Chancellor had ended his long deliberations and finally acted overshadowed the aspect of relations with Israel. Rather cautiously the paper agreed that 'by and large Erhard had made the right decisions', but expressed concern about the possible Arab reaction.[93] 'Nasser solved the problem of German–Israeli relations for us,' the paper wrote in a later article, 'and has forced us to take up this thread. Israel should respond without further conditions.' And it added the advice that the Federal Government should stick to its decision not to send arms to areas of tension.[94] Finally, the liberal weekly *Die Zeit* reminded its readers that 87 states had recognised Israel as well as the Arab states; some were exporting more arms to Israel than the Federal Republic and *Die Zeit* asks: 'Why do the Arabs measure with two types of yardstick?'[95]

By mid-May 1965 the depth of the crisis seemed to have passed, but for the West Germans much hard work remained to be done. The Federal Government still had to negotiate with the Israelis about the details of normalisation and the financial and economic aid that was to replace the arms deliveries. There was optimism that a satisfactory agreement would be reached. Much more intractable and unpredictable was the future of relations with Arab states, with whom the government now tried to mend its fences. The ending of West German development aid to Egypt was likely to be felt by the West German export industry, and there was talk in several Arab capitals of a boycott of West German goods. There was concern above all in Washington that the end of close West German–Arab relations would provide a further setback for the West since it would lead to an increase of Soviet influence in the Middle East. Immediately after the decision that Bonn sought diplomatic relations with Israel was announced and without waiting for Arab counter-measures, the Federal Government therefore instituted an intensive diplomatic campaign designed to try to repair the damage done to Arab–German relations. The plan, inspired by the SPD Bundestag member, Wischnewski, was to send special emissaries with good contacts in the Arab world to the Arab capitals to explain the position of the Federal Republic, to re-emphasise the West German desire for good relations with all the Arab states and, according to some commentators, to play down the importance of the new West German relationship with Israel. Further development aid in the form of generous offers of credits accompanied the efforts. All the men who travelled to Arab capitals had considerable political clout in the West German establishment, most also had important business connections.[96] One achievement of this campaign appears to have been that those Arab governments which broke off diplomatic relations with Bonn confined themselves to the closure of their embassies and expressed willingness to preserve their consular and economic relations, thus keeping open a life-line to the West German economy.[97] The formal rupture of diplomatic relations by the majority of Arab states could not, however, be prevented, nor were these re-established for several years. To the Federal Government it was of course most gratifying that its economic influence, the only influence it could wield, was not too seriously impaired and that no Arab state recognised the German Democratic Republic.

Apart from this temporary hiccup in German–Arab relations the crisis had no serious consequences and, as an international crisis, though unpleasant, cannot be regarded as a grave one. None of the four players involved in the dispute could register any really worthwhile gains or serious losses. The state that came off best was the GDR. It did not achieve international recognition but had at that stage, with hindsight, little chance of doing so. That its highest executive was invited officially to a Third World country was a useful consolation prize but certainly not unique, since a previous high East German personality had visited Cairo six years earlier. That such a visit was repeated at this time was, however, a sign of the East German state's increasing economic power and political importance, and of increasing Soviet influence in the Middle East generally. While the amount of development aid which Ulbricht had to offer was not particularly high and was not an element in Nasser's decision to invite him, it had one political significance that it enabled Nasser to defy the economic power and generosity of the Federal Republic. There may well have been some truth in Nasser's statement that West German aid was of limited advantage to him since, though it was much greater than anything the GDR could have afforded to offer, it was expensive in terms of the rate of interest the Egyptians had to pay. This is not to say that the Egyptian President could afford permanently to spurn any West German aid. But at this stage his priorities lay elsewhere; to help him in the pursuit of the civil war in Yemen and to shore up his tottering economy he needed both weapons and economic aid and these were available easily and cheaply in the Soviet Bloc. Hence his willingness to risk Bonn's displeasure.

It is in the field of his Arab policies that Nasser appears to have miscalculated. Firstly, it was doubtful whether he expected Bonn's reaction to be as fierce as it turned out to be. In view of Bonn's long-standing reticence as regards its relations with Israel, the Egyptian President may not have thought that the Federal Government would affront him by recognising the Jewish State. But once this had occurred he had to react strongly and needed to carry the other Arab states with him. This he was not able to achieve, either when he decided on recognition of the German Democratic Republic or when he broke off diplomatic relations with the Federal Republic. The price he paid in this crisis was more than anything the price of Arab unity. This was his most serious loss, a defeat – even if temporary – which the Israeli press did not fail to note, and for which the glamour of the Ulbricht visit and the diplomatic blow it struck at the West

123

was no compensation. Finally, the Egyptian government, together with those of other militant Arab states, lost a weapon they had wielded with some succes for nearly a decade: that of browbeating Bonn with threats of recognising the GDR. Nasser had been able to exert considerable influence on West German and indirectly thereby on western policies in the Middle East. Some Arab governments had also used it to press for more economic aid from the Federal Republic. While the possibility of exerting pressure on Bonn remained as long as the Federal Government clung to the Hallstein Doctrine, the normalisation of relations with Israel and Arab disunity had deprived the Arabs of credibility in the use of the doctrine as a political weapon. The Egyptian President's only prize therefore was his success in stopping the West German arms deliveries to Israel. With hindsight this now seems more of face-saving value, since it did not stop the supply of weapons to Israel and must be offset against the normali- sation of relations between Bonn and Jerusalem.

In a sense the Arabs' loss was Israel's gain. The Federal Republic's recognition of Israel after nine years of dithering was an Israeli victory scored over the Egyptians. That during these years Bonn entertained diplomatic relations with the Arabs while refusing them to Israel gave the Egyptian President an important propaganda point: it confirmed his 'thesis that Israel's presence in the region was an irregularlity'.[98] That Bonn had singled out Israel among the vast number of states with which it had normal relations was not only a humiliation for the Israelis but could also give the impression in the Arab world and elsewhere that the Federal Republic was in sympathy with the Arab aspirations and policies in the Middle East. The greatest gain for Israel in the medium term was that it could establish through Bonn a new formal link with western Europe and the European Community in which the Federal Republic was playing an increasingly important role. This was essential at the time for two reasons: firstly, that Israel's relations with the United States were beginning to cool, partly due to the latter's increasing preoccupation in Vietnam, partly because of the difficulties of American relations with the Arabs; and secondly, because relations with France, hitherto Israel's closest ally in Europe were beginning to change as France's colonial rule in North Africa was coming to an end and de Gaulle's government was trying to reassert French influence in the Mediterranean by improving his country's relations with the Arab world. Against these gains Israel had lost the West German arms supplies. This can be seen in retrospect as a minor diplomatic defeat, since it gave a moral victory to Israel's opponents and appeared at first to the Israelis as yet another failure by the Germans to live up to their image of rueful pentitents for sins committed in the past. But in practice the West German arms sources were replaced by others and Israel's successes two years later showed that the country's security had not been seriously impaired.

The country most adversely affected by the crisis appears at first sight to be the Federal Republic, though by comparison with other international disasters[99] the loss suffered by Bonn should not be exaggerated. But the Federal Republic had been at the centre of the crisis. It had tried actively to court two sides locked in a deep and bitter conflict, giving both economic and military aid to one side while to the other side it had also given economic aid and looked on while West German nationals participated in that side's war effort. Such support given to two mutually hostile parties, if continued over a long period is always dangerous. Even so it is not uncommon for states to give military aid to two sides of a conflict. The weakness in the case of Bonn was that it was pursuing its own diplomatic battles quite outside the turmoils of the Middle East. The Hallstein Doctrine, effective in the late 1950s in upholding the West German claim to be the sole representative of all Germans, had lost its force as a result of the strengthening of the Communist Bloc which, because of its rapid strides in nuclear and rocket technology, had created a strategic balance with the West. This increase in Soviet power had manifested itself in the Third World by greater political influence, especially in the Middle East. The increased economic strength of the GDR in the 1960s had allowed this state to make its own modest but not insignficant impact in the Third World.

The result was a diplomatic and, some said, a national humiliation for the Federal Republic. Undoubtedly clumsiness in managing the crisis and slowness in taking decisions were aggravating factors. The most persistent criticism levelled at the Federal Government by politicians and the media was that after Nasser's representations to Gerstenmaier in November 1964 the government took no action instead of immediately stopping the arms deliveries to Israel. But it is difficult to see how Bonn, under pressure from both the Americans and Israelis, could have acted otherwise or how, if it had, it would have changed the sequence of events. For it is clear that the invitation to Ulbricht to visit Cairo was not the result of the arms deal. Once the visit had been announced it was certainly foolish to expect Nasser to withdraw the invitation or to try to bribe other Arab states publicly to distance themselves from Egypt; the Spanish mediation may have been organised in undue haste, but the only alternative, to ignore the visit altogether, would have meant the coup de grâce for the Hallstein Doctrine and would have been regarded as a defeat for the Federal Republic. On the other hand Nasser's demand for an end to the arms deliveries to Israel had to be taken seriously. Whatever the importance of these deliveries may have been, their emotional impact in

125

the Arab world was too great and the risk of uniting the Arab states in favour of recognising the GDR too dangerous. Even the recognition of Israel carried such a risk; that the Federal Government in this instance decided to call Nasser's bluff is a positive factor in the crisis and to the credit of the Chancellor. Nor is it likely that greater support from West Germany's allies would have saved the Federal Republic from defeat. The Americans had not at first been a party to the secret arms agreement, though they had been informed about West German military aid to Israel around the time of the Adenauer–Ben-Gurion meeting in 1960. They did not become linked with the affair until shortly before the secret was leaked in 1964. They were torn, like the West Germans, between supporting Israel and the Arabs but had been avoiding military aid to Israel. They did not have the burden of a Hallstein Doctrine, indeed they were now opposed to the West German policy of the political reunification of Germany, which was no longer in line with their own policy of super-power détente. Intervention in the Ulbricht dispute would not have furthered this policy and, besides being ineffective would have drawn upon them more anger from the Arabs. Nevertheless the US did eventually at least save Bonn from deep embarrassment in its relations with Israel by taking over the arms supplies. All of which seems to confirm that the Federal Republic, having once manoeuvred itself into a crisis by the complexity of its own policies, had little chance of avoiding a humiliation of some sort.

Luckily for the West Germans, the crisis of 1965 left no serious scars for them or for their government. The split in the Erhard coalition between the Foreign Minister and the FDP on the one hand and the remaining CDU ministers on the other, which at one point had threatened a cabinet crisis, was soon healed. In the general election in October of that year the Erhard government was returned with a slightly increased majority. But as complaints about the Chancellor's conduct of other matters increased and an economic crisis, totally unconnected with Middle Eastern affairs, blew up the following year, the public were undoubtedly reminded of the Middle Eastern debacle and Erhard's failure at times to provide strong leadership. To that extent his handling of affairs in 1965 contributed to his downfall in the autumn of 1966.[100] In the domain of West German foreign policy the crisis accelerated a trend which had already started: the gradual weakening of the Hallstein Doctrine and in general, the policy of reunification through strength. Opposition within the Federal Republic to this policy had developed in the early 1960s when it became clear that changes in the International Political System were working against the reunification of Germany. The debacle of West German foreign policy in the Middle East now showed that the Hallstein Doctrine not only was no

126

longer an efficient tool to keep the GDR ostracised but could be dangerous in the hands of the Federal Republic's opponents as an instrument of blackmail. As a result changes were soon made in the direction of West German foreign policy and these had some effect on the relationship with Israel.

NOTES

1. *Frankfurter Rundschau*, 26 Oct. 1964.
2. G. von Hase, 26 Oct. 1964, dpa report quoted by Seelbach, op. cit., p. 123.
3. Ibid., 30 Oct. 1964, ibid.
4. Meaning: insufficient efforts (author's comment).
5. See Shinnar, op. cit., p. 146.
6. Ibid.
7. *Israel Digest*, 23 Oct. 1964.
8. *Kessing's Contemporary Archives*, pp. 20741–2 (1964–65).
9. *Neue Zürcher Zeitung*, 17 Jan. 1965.
10. Khrushchev, shortly before his dismissal, had promised Egypt new arms deliveries.
11. H-H. Gaebel, 'Unangenehme Möglichkeiten' in *Frankfurter Rundschau*, 3 Oct. 1964.
12. *Neue Zürcher Zeitung*, 1 Nov. 1964.
13. *Frankfurter Rundschau*, 3 Oct. 1964.
14. See also Arnold Hottinger: 'Background to the Invitation of Ulbricht to Cairo' in *Europa-Archiv*, 1965, no. 4.
15. A. Hottinger, op. cit.
16. *Der Spiegel*, 24 Feb. 1965, p. 25. Sometimes quoted as $100 million.
17. O. Frei: 'Foreign Political Exertions of the GDR in the Non-communist World', *Europa Archiv*, 1965, no. 22.
18. A. Hottinger, op. cit.
19. *The Times*, 16 Jan. 1965.
20. A. Hottinger, op. cit.
21. *Al Ahram*, quoted by the *Egyptian Mail*, 6 Feb. 1965.
22. *Al Ahram*, quoted by *The Times*, 8 Feb. 1965. The *Al Ahram* reports on the Nasser–Federer talks were criticised by Bonn as 'tendentious'.
23. *Europa Archiv*, 1965, no. 15 (diary of events).
24. *FAZ*, 3 Feb. 1965.
25. *Frankfurter Rundschau*, 8 Feb. 1965.
26. *Europa Archiv*, 1965, no. 15.
27. *FAZ*, 28 Jan. 1965.
28. *FAZ*, 30 Jan. 1965.
29. *Die Welt*, 28 Jan. 1965.
30. *Die Zeit*, 5 Feb. 1965.
31. *Handelsblatt*, 4 Feb. 1965.
32. *Rheinischer Merkur*, 4 Feb. 1965.
33. *Neue Rheinzeitung*, 2 Feb. 1965.
34. Report quoting de Nerva in *Le Monde*, 12 Feb. 1965.
35. *Jerusalem Post*, 16 Feb. 1965.
36. *Jerusalem Post*, 8 Feb. 1965.
37. *Al Hamishmar*, 7 Feb. 1965.
38. The seat of the East German government.
39. *Jerusalem Post*, 12 Feb. 1965.
40. *Davar*, 14 Feb. 1965.
41. *Jerusalem Post*, 16 Feb. 1965.

42. *Bundestagsprotekolle*, IV/164, 17 Feb. 1965, p. 8103.
43. *Jerusalem Post*, 19 Feb. 1965.
44. *New York Times*, 18 Feb. 1965.
45. *The Observer*, 14 Feb. 1965.
46. *The Observer Foreign News Service*, 16 Feb. 1965.
47. *New York Times*, 16 Feb. 1965.
48. Quoted by *Der Spiegel*, 24 Feb. 1965.
49. *Daily Telegraph*, 16 Feb. 1965.
50. Quoted by *The Times*, 13 Feb. 1965.
51. *Die Zeit*, 19 Feb. 1965.
52. *Abendpost*, 16 Feb. 1965.
53. *Handelsblatt*, 16 Feb. 1965.
54. *Stuttgarter Zeitung*, 16 Feb. 1965.
55. *Die Welt*, 16 Feb. 1965.
56. *FAZ*, 16 Feb. 1965.
57. *FAZ*, 13 Feb. 1965.
58. Ibid., 12 Feb. 1965.
59. *Der Spiegel*, 24 Feb. 1965.
60. *New York Times*, 25 Feb. 1965.
61. Seelbach, op. cit., p. 252.
62. *FAZ*, 8 March 1965.
63. Birrenbach: 'The Establishment of Diplomatic Relations between the Federal Republic of Germany and Israel' in *Ludwig Erhard, Contributions to his Political Biography*, 1972.
64. *Deutschkron*, op. cit., p. 313.
65. Birrenbach, op. cit.
66. Ibid.
67. Rainer Barzel in: *Die politische Meinung*, May 1991.
68. *Deutschkron*, op. cit., p. 315.
69. *Bulletin of the Press and Information Office of the Federal Government*, Bonn, 1965/41, 9 March 1965.
70. Shinnar, op. cit., p. 130.
71. Shinnar, op. cit., p. 131.
72. *The Times*, 3 March 1965.
73. Birrenbach, op. cit.
74. Letter, Erhard to Eshkol, 12 May 1965 in *Bulletin* 1965/84, 14 May 1965.
75. See chapter 6.
76. Birrenbach, op. cit.
77. *Jerusalem Post*, 8 March 1965.
78. *Davar*, 9 March 1965.
79. *Davar*, 14 March 1965.
80. *Haboker*, 9 March 1965.
81. *Ha'aretz*, 10 March 1965.
82. *Hatsofe*, 11 March 1965.
83. *Herut*, 11 March 1965.
84. *Jerusalem Post*, 17 March 1965.
85. *Davar*, 17 March 1965.
86. Seelbach, op. cit., pp. 115–116.
87. *FAZ*, 15 March 1965.
88. *Die Welt*, 8 March 1965.
89. '*Die Welt*', 16 March 1965.
90. *Frankfurter Rundschau*, 9 March 1965.
91. *Südd. Ztg.*, 9 March 1965.
92. *Südd. Ztg.*, 10 March 1965.
93. *FAZ*, 11 March 1965.
94. Ibid., 12 March 1965.

95. *Die Zeit*, 19 March 1965.
96. Seelbach, op. cit., pp. 139–41.
97. *Die Welt*, 12 April 1965.
98. Eliezer Livne: 'Collapse of a Policy' in *New Outlook*, vol. 8, no. 2, Tel Aviv, March 1965.
99. E.g. the consequences for the UK of the Suez Crisis of 1956.
100. Cf. Besson, op. cit., p. 326.

8

Normalisation – the Conflict between the Moral Debt and Political Realism

The establishment of diplomatic relations between the Federal Republic and Israel – the official date was 12 May 1965, the day the letters were exchanged between Chancellor Erhard and Prime Minister Eshkol – and the near-certainty that the outstanding issues between the two countries would be solved, did not at first create complete harmony. There was initial disagreement when it came to the choice of the two ambassadors. The negotiations about West German economic aid to Israel to replace the expiring Luxembourg Restitution Treaty, foreshadowed in Erhard's letter to Eshkol, led to serious differences of principle between the two governments. As so often, Israeli suspicions arising from the past complicated matters. These were not assuaged by a new, albeit temporary political development in the Federal Republic; the sudden, quixotic rise to prominence of neo-Nazism, momentarily reawakening all over the world memories of the rise of the Nazis to power in Germany in the 1920s and early 1930s. These difficulties were, however, ultimately overcome and relations improved. The establishment of diplomatic missions never guarantees that relations are 'good', even less that they are 'special'. They had been 'special' in the case of the Federal Republic and Israel ever since their commencement in the early 1950s. The question now was whether 'normalisation', apart from the exchange of diplomatic missions, also meant that the relations would maintain their special character or become indistinguishable from those between other countries.

The man chosen by the Federal Government to be the first West German ambassador to Israel was Dr Rolf Pauls, a career diplomat who, the Germans asserted and the Israelis acknowledged, had never been a member of any Nazi organisation, but had fought as an officer of the Wehrmacht, the regular army, on the Russian front during the Second World War, had lost an arm and been awarded the *Ritterkreuz*,[1] the highest German award for bravery. This fact made him unpopular with the Israeli public, some of whom, especially those belonging to the organisations of former concentration camp victims, engaged in strong, sometimes violent demonstrations. Most of the atrocities against Jews in German-occupied areas were perpetrated by the SS, Gestapo and ancillary organisations but the Wehrmacht too, was subject to Hitler's personal orders and not free from blame. Some of those opposed to Pauls' appointment also argued that fighting for Germany in the Second World War was a way of supporting the Hitler regime. Even the Israeli government and the more moderate elements of the public would have preferred either a person who had proved his friendship to Israel, such as Professor Boehm or Dr Birrenbach or, failing this – and both these men are said to have refused to be candidates for this post[2] – a man too young to have been associated with the Third Reich. At best, therefore, Dr Pauls' selection was regarded as unimaginative. The Israeli authorities, for practical reasons, were prepared to accept Pauls as West German ambassador. To them it sufficed that he was free from the burden of Nazism; what seemed more important was that he was well established in political circles in Bonn and that he would therefore be listened to by and have influence with the people who mattered. They must have realised that the exclusion of members of the former German fighting forces would have reduced the choice for this difficult post to an unacceptable level. In any case they could not have rejected Pauls, especially once the German reaction to the Israeli attitude became apparent, without putting the whole future of their relations with the Federal Republic at serious risk again. Even so, the Israeli government could not totally ignore the pressure of public opinion, and while it tried to persuade the West German authorities to think again, there was some delay in Pauls' appointment being approved by Jerusalem.

It must be assumed that the Federal Government, eager to normalise relations with the Jewish State and aware of the delicacy of the task, had carefully considered Pauls' candidature. Accusations in the Federal Republic and Israel that the choice of Pauls by the Foreign Office was an act of vindictiveness for having been overruled in the matter of establishing full diplomatic relations with Israel, or that it was influenced by Arabists who wished to cast new clouds on West German–Israeli relations have been largely discounted. To the Germans it did not seem objectionable that a person should have fought for his country, even under Hitler. Clear distinctions were drawn between being a soldier in wartime, a duty that no able-bodied man could escape, and being an active supporter of the Hitler regime through the Nazi Party or its organisations, or by holding a prominent public position. To identify ordinary war service in the fighting forces with support for the regime would be anathema to all Germans

131

and would stigmatise most of those who had a clear non-Nazi or anti-Nazi record. That would have included Chancellor Erhard himself and many of his cabinet. This explains the anger expressed in the Federal Republic over the Israeli attitude in regard to Pauls. Erhard's personal reaction, to which he gave expression at a press conference, was that he saw no reason to discriminate against anyone because he had been a German soldier doing his duty.[3] So the Federal Government remained firm.

Pauls was not unknown in Israel. He had been an official at the West German Foreign Office, dealing with questions of development aid and in this capacity had participated in the last stages of the Birrenbach negotiations, during which the question of future economic aid to Israel was discussed. On 1 July Jerusalem finally gave its agreement for Pauls to take up his post, but his troubles were only just beginning. A more difficult and delicate task than the one he faced as the first West German ambassador to Israel when he arrived there in August 1965 can hardly be imagined. As one Israeli author put it:

> With all the goodwill of those present the fact could not be obliterated that Pauls represented a nation in whose name unheard-of crimes had been committed. That twenty years later they had to shake hands with an official representative of this nation, however little he himself may have had to do with those crimes, was indeed difficult .... Hardly any German diplomat had ever been in a comparable position.[4]

This was the situation when on 19 August Pauls handed his accreditation papers to President Shazar of Israel. At the ceremony, while demonstrators were shouting abuse outside, Pauls told his Israeli audience inside the presidential residence that 'the new Germany looks back with sorrow and revulsion to the horrendous crimes of the National Socialist regime'. Since then, he continued, people of goodwill on both sides had worked patiently to pave the way for this new beginning of the relations between the two peoples. 'We hope that the exchange of ambassadors will contribute to successful progress along this path.' Shazar, alluding to the Holocaust, replied *inter alia*:

> The presentation of these credentials is proof that chaos does not last for ever and that even the worst night is followed by a dawn. It is our duty to concentrate our energies to stop forever the progression of hate, and may the spirit of that accursed time never be born again.[5]

The President then assured Pauls that he would receive all the support he needed for his task.[6]

The difficulties over the appointment of Pauls illustrate once again the peculiar relationship then still existing between the two countries. It is not unheard of that the host country, when it is to receive a new ambassador from another state, raises objections on political or personal grounds. The matter is then usually settled either by the host country accepting the person designated by the other state or by another person being sent. In the case of Pauls there were no justifiable personal or political reasons for refusing him. He had not made himself unpopular either with the Israeli government or public by saying or doing anything that could have been offensive in their eyes. Nor could his presence in Israel have been politically harmful to Israel in any way. That he was eventually appointed was due to the fact that both governments based their decision on practical considerations and rejected the irrational criticism in both countries, more especially the emotional reactions of sections of the Israeli public. It was clear from the start that Pauls, or any other person if the West Germans had given way, would be closely watched in Israel. It was not Pauls the man but the German people he represented who would be continually on trial; this showed that despite diplomatic relations little changed at first from the Israeli side and that normalisation in the general sense was still some distance away.

The tension created in Bonn and Jerusalem by Pauls' appointment was further increased by another appointment made by the West German Foreign Office and to which many Israelis also took exception. On 8 August, three days before Pauls, the West German chargé d'affaires and counsellor of the new embassy, Alexander Török, a naturalised German of Hungarian origin, arrived in Israel. He was second in the hierarchy of the embassy, that is he ranked after the ambassador himself and was also his deputy.

Török had been sent by Admiral Horthy, the Regent of Hungary, to the Hungarian embassy in Bucharest as attaché in 1940. Four years later, with the Third Reich in virtual control over Hungary, he became secretary at the embassy in Berlin. The function of the Hungarian embassy in Berlin is said by 1944 to have been mainly to receive and pass on the orders of the German government to Hungary, including that of the extermination of Hungarian Jewry. Török's case was therefore much more suspect and his innocence more difficult to prove than that of Pauls'.[7] When Török's appointment became known there was a new outcry in Israel, where he was accused of participating in the extermination of Hungarian Jews. But enquiries in both the Federal Republic and Israel failed to pin any involvement in the persecution of Jews on him. Israeli organisations which gathered material about the persecution of Jews in the Third Reich and states allied to it had no knowledge of him.[8] As a result his appointment was approved by the Israeli Foreign Minister and the storm eventually died down, but not before considerable anger had been expressed by the Israeli press, supported on this occasion by some West German newspapers. Once again the question was asked whether in choosing Török the Federal Foreign Office had been motivated by resentment at diplomatic relations being established against its wishes.[9]

The appointment, proposed by the Israeli government, of Asher Ben-Natan as ambassador in Bonn ran into difficulties in the Federal Republic. Ben-Natan had been a high official at the Israeli Ministry of Defence concerned with the arms deliveries by the Federal Republic to Israel under the secret arms agreement. His appointment was therefore regarded by some West German politicians, especially by those concerned about the effect it might have on the Arab states, as an error of judgment by the Israeli government. In Bonn, too, the authorities felt uncomfortable and the government delayed giving its approval. The problem was not made easier by rumours, circulating in some Arab countries and echoed by the East Germans ever since the Birrenbach mission had been negotiating in Jerusalem, that Birrenbach had countenanced a new secret military aid agreement with Israel.[10] The proposed appointment of a man so deeply involved with the former arms deliveries could, it was felt, increase Arab suspicions further. That such a new agreement existed was strongly denied by the Federal Government spokesman, von Hase, and described by Erhard as slander.[11] In fact there is no possibility after the recent diplomatic debacle and the strong criticism in the Federal Republic of the arms agreement, that a new arms deal could have been contemplated by the Federal Government. As in the case of the Israeli government in respect of Pauls, Bonn in the end overcame its qualms after some delay and gave its agreement to Ben-Natan's appointment.

More serious than the differences over the appointment of the diplomats were those over the economic aid the Federal Republic was to grant to Israel. In his letter to Eshkol dated 12 May, Erhard had stated that 'his government was ready to enter into negotiations with the Israeli government about future economic aid in two to three months' time'.[12] But the whole problem of West German economic aid to Israel was burdened by a legacy of vagueness and uncertainty bequeathed to the current two governments by the Adenauer–Ben-Gurion meeting of March 1960 in New York. Only the scantiest information came out of that meeting, the gist of which was that Ben-Gurion had proposed that Bonn might participate in development projects for the Negev desert and that Adenauer had given his agreement in principle, saying that West Germany would help Israel 'for moral reasons and for reasons of logic'.[13] The amount involved is believed to have been $50 million per annum over 12 years in the form of a loan. The agreement was honoured but, like the arms deliveries discussed in equally vague terms, never published. Neither was it ever passed by the Bundestag as would normally have been the practice, no doubt for fear of Arab reactions. Another factor which came into the negotiations about to begin between the two countries in the summer of 1965 was that the Restitution Agreement concluded in Luxembourg in 1952 was due to expire in 1966. A letter by the then Israeli Foreign Minister, Moshe Sharett, written on the day of the signing of the agreement, had included a paragraph saying that once all its requirements had been met by the Federal Republic, no further claims on it for restitution would be made by Israel. Ben-Gurion was reported to have told Adenauer in New York that Israel could not demand more restitution as a matter of right, but that he hoped that the West Germans would agree to it. If the reports of Ben-Gurion's statements to Adenauer are to be believed, in which the Israeli premier returned to the question of the depopulation of the Jewish people and the resulting difficulties of defending and developing Israel,[14] then there seems little doubt that he was hoping Adenauer would agree to continuing the process of restitution for Nazi crimes against the Jews. Adenauer's agreement and reference to 'moral' as well as 'logical' (that is, political) reasons could be taken to mean that he too saw it that way. It certainly appears to have been the view of Eshkol's government when, at the end of 1965, later than had been proposed in Erhard's letter, the Israelis were negotiating on how to make a new economic aid agreement which would incorporate what had been agreed by Ben-Gurion and Adenauer five years previously. The Federal Government, however, no doubt wary after having been caught over the similarly vague arms agreement, no longer took that view.

In common with most other developed countries the Federal Republic was giving substantial economic aid to developing countries in the 1960s. By 1961 economic aid had become a sufficiently important factor in West German economic and foreign policy to justify the establishment of a separate government department, the Ministry of Economic Co-operation, in Bonn. It was put in the charge of Walter Scheel, a Free Democrat, who was to play an important foreign policy role in the SPD–FDP coalition led by Willy Brandt, and who later still served a term as West Germany's Federal President. Much of the economic aid was given in the form of low interest loans. Two basic principles were applied: firstly, that the money should be used for specific projects in the recipient countries and that its use on these projects should be approved by the Federal Government; and secondly, and perhaps less rigorously, that it should not be given to states which actively opposed the West German policy of reunification and non-recognition of the GDR. These two principles were reiterated by

Erhard in a government statement issued at about the time when preliminary discussions between his government and the Israelis on West German
aid to Israel were starting. He said that development aid by the Federal
Republic would be 'measured by whether it made sense in the recipient
country and whether the recipient country recognised the principle of
self-determination and national unity which Germany claimed for itself'.[15]
The economic aid promised to Israel in Erhard's letter was to be of that
kind. The question of German reunification was not important in this case
as it was out of the question that Israel would recognise the GDR, though
an incident did occur that offended the Federal Government. But Israel
objected strongly to the conditions that the money should be used for
specific projects and that its use should in each case have to be investigated
and approved by the German authorities. This was what was meant when
the Minister for Economic Development, Walter Scheel, told Rolf Vogel,
editor of *Deutschland-Berichte* that the West German–Israeli negotiations
would be about aid that was 'project bound'.[16] The Israeli objection was
that this was a departure from the principles adopted by Adenauer and
Ben-Gurion in New York, while the Germans claimed that Adenauer had
simply enunciated the principle of continuing West German economic aid
to Israel and that no details had been discussed at the time.

The truth, as so often in such cases, lies somewhere in between. It had
been stipulated by Ben-Gurion at the New York meeting that the money
would be used for development in the Negev desert, and in the months
that followed the meeting it had been agreed that it would take the form
of 'commercial loans' amounting annually to approximately DM150–200
million. Since Bundestag approval could not be obtained because of the
need for secrecy, it could not be incorporated into the West German
development aid programme. For this reason also the agreement was, as
Shinnar put it, morally but not legally binding. Shinnar makes it clear,
however, that Adenauer had demanded a detailed plan for the development of the Negev, which was to have formed the basis for the negotiations.
He also mentions that at regular intervals the Israeli mission in Cologne
was to furnish the bank which issued the loans with details of how amounts
were used within the framework of the Negev development plan.[17] Up to
the time when the negotiations for a new agreement were beginning in
1965, 560 million of the DM2,000 million originally agreed had been
paid.[18] Both sides had accepted that the time for secrecy was now past. For
the West Germans this clearly meant that this new agreement, like all
other development aid agreements, should be concluded along the general
lines of German development policy.

From the practical point of view the Israelis may have feared during
the period of negotiations that the German conditions might mean smaller

loans. But the amounts finally agreed, DM160 million per annum for a period of 25 years at no more than 3 per cent interest did not fall short of the amounts granted under the Adenauer–Ben-Gurion agreement. From utterances by Israeli politicians it is also clear that the reasons for their objections were not of a practical nature. They did not want Israel to be treated in the same way as the other developing countries. They wanted the continuation of *shilumim*,[19] that is they wished West German aid to continue as a mark of the Germans' moral debt for the crimes committed against the Jewish people during the Second World War. This view was expressed by the chief Israeli negotiator, the Israeli ambassador in Bonn, Asher Ben-Natan:

> I believe that there is a kind of restitution which, from the German side, can never come to an end. In this sense restitution does not necessarily stand for money. What I mean is that restitution should be and must be an inner need, a fundamental attitude and above all a willingness of a special character.[20]

An agreement was finally signed on 12 May 1966. It followed the West German rather than the Israeli approach. The loans were to be financed, as were those negotiated secretly following the Adenauer–Ben-Gurion meeting, by the Kreditanstalt für Wiederaufbau in Frankfurt. Much of the aid was to go to housing, about 25–30 per cent to finance small and medium-sized businesses. A smaller amount went to the Negev. The agreement paved the way for further improvements in the relations between the two countries, which were nevertheless disturbed by a number of other events occurring at that time. One incident, foreshadowed earlier in this chapter, concerned West German relations with the Soviet Bloc. Ironically, at about the time of the signing of the agreement, the Israeli government came out strongly in favour of a recognition of the Oder–Neisse line as Germany's permanent eastern frontier. This was almost as sensitive a point to the Federal Government as recognition of the GDR. At the Potsdam Agreement of July 1945 the German territories east of this line were separated from Germany and subsequently occupied mainly by Poland, but a small area by the USSR. Interpretation of the agreement had differed: the Soviet Bloc claimed that these areas now belonged permanently to the two countries that occupied them, the Federal Republic supported by the West took the line that they had only fallen under Polish and Soviet administration and that their fate would be decided by a peace treaty with Germany. Successive Federal Governments had therefore refused to recognise the Oder–Neisse line as the permanent frontier of Germany in the east. The Israeli call for recognition, which may have been prompted by the parallel with Israel's own

unrecognised and insecure borders or, more rationally – with a view to
the negotiations about economic aid – designed to remind the Germans
that they could not take Israel for granted, caused anger in the Federal
Republic at the worst possible moment; so much so that Israeli ambassador
Ben-Natan found it necessary to express regret that the statement about
the Oder–Neisse line should have coincided with the negotiations about
the aid agreement.[21] As it happened, the agreement was not affected by
this incident.

Two other events, speeches in fact, by prominent personalities in West
German–Israeli affairs, should be mentioned since they occurred at and
have relevance to this period of adjustment, when the two states sought in
different ways to achieve normalisation in their relations. The first was
the speech by the Israeli Prime Minister, Levi Eshkol, when he received
in his home former Chancellor Adenauer, who was nearing the end of his
first visit to Israel. Adenauer, who had not been able to come to Israel
during his term of office, was received by the Israelis as a very honoured
guest, and although there were a few derogatory comments in certain
sections of the press, it was clear that his moral stance towards the Jewish
people and the help he caused to be given to Israel were by now fully
recognised and appreciated by a majority of Israelis and by the government. In this context Eshkol's after-dinner speech may seem surprising
and certainly caused embarrassment or even shock to his guest. Eshkol
spent a great deal of time detailing in rather strong language the atrocities
committed against the Jews by the Nazi regime and reiterated with
emphasis the losses of the creative powers which his people had sustained
in the fields of science and culture as a result of the attempted extermination. In a reference to the rebuilding of their homeland by the Jewish
people he quoted the words of the prophet: 'In thy blood thou shalt live.'[22]
Only after that did he mention Adenauer as a man who had 'reflected
deeply on the unique spiritual character of Israel and the significance of
the survival of the Jewish people through the millennia', reminding him,
however, that the Restitution Agreement that he, Adenauer, had signed
in the name of the German people was no atonement as there was no
atonement for annihilation.[23]

Eshkol may have thought that this was a necessary admonition that
could not be uttered often enough for the benefit of the millions all over
the world who were ignorant of or indifferent to what had occurred during
the Second World War. Or it may have been meant for internal consumption.[24] Adenauer, on the other hand, could be excused for thinking
that he needed no reminder and clearly felt angry that such a speech was
made at a gathering at which he was the honoured guest. In his reply,
during which he was said to have shown some irritation, he reminded his

host that the Germans had done everything and given every proof that 'we intend to overcome this period of horror, which cannot be undone'. But he added that it should now be left to history. 'I know how difficult it is for the Jewish people to accept this. But if goodwill is not recognised then no good can come of it.' He added in a reference to the building of the Jewish State:

> You have a wonderful task, to show to all countries the way to exemplary achievement in rebuilding your land to which all who wish can return. Could you not find in this your new task some consolation for what has been done to you and which neither I nor anyone else would want to excuse? This is a fact one should never forget.[25]

Quite clearly this turn of events during one of the highlights of Adenauer's visit was unexpected and added a sour note. Adenauer later wrote:

> We have a claim, however great our guilt vis à vis the sorely tried Jewish people, that our good intentions be recognised .... Nothing hurts more deeply than suspicion and rejection of goodwill.[26]

Eshkol's speech was criticised in Israel and he may have realised that he committed an error of judgment for he arranged a 'placatory chat' with Adenauer just before the latter's departure for Germany.[27]

Adenauer's injunction to the Israelis to draw the line under the past and take a more positive approach to West German–Israeli relations was reiterated by the new West German ambassador, Rolf Pauls, in a speech made on the occasion of the Israeli Industries Fair in June 1966 at which the Federal Republic had for the first time opened its own pavilion. Like Adenauer, Pauls praised the work of building up a new Jewish State and dwelt at some length on the difficulties of establishing new relations with the Jewish people and with Israel after the events of the 1940s: for this, infinite patience was needed, he added. 'It is unavoidable,' he stressed, 'that in our relations the past overshadows the future more than is the case anywhere else. But we should not thereby allow ourselves to be robbed of all confidence in the future.' He then added:

> It is with great concern that we observe how for reasons of political advantage or out of selfish motives the sufferings of the past are constantly stirred up in order to disturb the present so that it cannot serve the future. These forces which are hostile to present-day Germany do a bad service to their own country by their agitation.

Pauls complained particularly about some Israeli news reporting which, he said, depicted the West Germany of 1966 as a Nazi state, something that no longer existed and would never exist again. In a reference to the rise of the neo-Nazi NPD during 1966 he said that to regard fringe developments as symptomatic was bad reporting. He deprecated recent comments in Israel that Germany needed to show by deeds that it was worthy of being received back into the international community. He said that the Federal Republic was already occupying a respected place in the community of nations and needed no further dispensation. He also made the point that the quality of German–Israeli relations would be fundamentally affected by the way Israel treated German interests – a reference to Eshkol's declaration on the Oder–Neisse line. Pauls ended by saying that relations with Germany would improve to the extent to which policies were conceived not out of the past but with a view to the future.[28] This speech caused a severe reaction in Israel where Pauls was criticised for being arrogant. Some editorials called his speech 'a mixture of preachment and stricture'.[29] But his government supported him and when leaving for Bonn on a brief visit he made it clear that in relations between Israel and West Germany criticism was not the unilateral right of Israel; the Israelis would just have to get used to that fact.[30] To be fair, these strictures all occurred in the second half of the speech, while the first half reiterated the self-criticism which characterised so many statements made by West Germans in regard to Israel. Perhaps precisely for that reason the Israelis found the latter part new, almost sensational and shocking. Whatever the justification for some of Pauls' criticisms, they were not calculated to improve matters in an area where the 'normalisation process' was lagging behind: in the relations between the West Germans and the Israeli public, as compared with the Israeli government. As the *Neue Zürcher Zeitung* Tel Aviv correspondent put it: perhaps Pauls could have used the 'infinite patience' he had preached in the first part of his speech.[31] His utterances did, however, reflect an increasing self-assurance on the part of the West Germans, which was noted at the time in the Federal Republic's foreign relations generally. It was a message to the Israelis that the one-sidedness of the relationship was over, that the Germans expected something in return for their generosity, something rather more positive than declarations about a 'firm' Oder–Neisse frontier. The message was received in some Israeli circles. The sharp reaction to the speech by the Israeli media and public was in turn criticised by Israeli officials. Ben-Natan, referring to Pauls' point about the Federal Republic's international respectability, said that the civilised world had not empowered Israel to speak in its name.[32] Furthermore, the Israeli Foreign Minister, Golda Meir, reiterated Ben-Gurion's often repeated assertion that the Germany of today was not that of the Nazi period; if the Israelis did not accept that as a fact, then the defeat of Hitler would be meaningless.[33]

All the events discussed so far in this chapter, the appointment of ambassadors, the difficulties over economic aid and the speeches by Eshkol, Adenauer and Pauls, indicate a new twist in the German–Israeli relationship. Although diplomatic relations had finally been established and the disagreements over the West German scientists in Egypt and the Statute of Limitations had faded into the background, a new fundamental disagreement loomed over the development of relations between the two countries. When looked at closely it was not really quite new. The Israelis had from an early date taken the line that any single act of restitution did not repay the debt which the Germans owed them as a result of Hitler's actions. Theirs was an eternal debt which no aid or support could expunge. They were therefore expected to continue to support Israel as strongly as was possible for an indefinite period. That, basically, was behind one of the arguments of both Ben-Gurion during his meeting with Adenauer in 1960 and Eshkol during Adenauer's visit in 1966: that the loss to the Jews of creative powers during the Holocaust was an ongoing handicap to the development of the Jewish State. It could be true both for economic development and for defence; hence the request for arms in the late 1950s and early 1960s. The important point in 1965–66 was that the Luxembourg Agreement of 1952 was about to expire and that at that time the Israeli government had acknowledged that once the Germans had carried it out the Israelis had no further claim on them. The Jerusalem government was now at pains to persuade the West Germans that this did not mean that they had been released from their moral debt.

Adenauer, after his return from Israel, and despite the brush-off he received from Eshkol, still accepted the Israeli view, as he had done in the past. He wrote *inter alia*:

> My impressions and experience in Israel have strengthened my conviction that we must never desert this struggling state .... To those who think restitution must come to an end and that the German people cannot condemn itself to the slavery of eternal guilt I want to say that one cannot put figures to a moral obligation nor pay it off penny by penny. What we shall do in future to aid Israel ... should come more and more under the heading of cooperation leading to reconciliation.[34]

But the German view was changing. More people in the Federal Republic were becoming impatient with frequent admonitions about Germany's past, made both inside and outside Germany. To many, 'normalisation' of Germany's relations with Israel meant that, although one must not forget the past, these relations should now be the same as with all other states. Worldwide détente, though not as yet extended to the Federal

Republic in its relations with the Soviet Bloc, nevertheless removed some of the fears of a major world war and therefore increased West Germany's independence and self-assurance in world affairs, as did also the West Germans' economic strength and the successes of the Federal Republic in its European policies. If these trends in international politics continued they could have the effect of further stiffening West German resistance to Israeli demands for unqualified support.

The first signals that changes in the Federal Republic were on the way came at the end of 1966. A temporary economic recession, the deepest since the War, occurred worldwide. It was far from being as serious as in the 1930s or more recently in the 1970s and 1980s, but the Federal Republic was affected more severely than the rest of the industrial world. Suddenly, in the country of the economic miracle, there were over a million unemployed and exports, on which the West Germans' recent prosperity had mainly depended, were declining. The government of Ludwig Erhard, after its election victory only a year earlier, fell over disagreement on what to do with the budget deficit. As a result a Grand Coalition consisting of the two largest parties in the Bundestag, the CDU/CSU and the SPD, was formed while the small FDP, the junior coalition partner in all but one of the previous governments, went into opposition, alone, with only 49 members. This coalition, unique in the history of the Federal Republic, was regarded as unsatisfactory since the absence of a viable opposition was seen as contrary to the principle of sound parliamentary democracy. This added a political dimension to the already acute uncertainty caused by the economic recession. Both recalled to memory similar events occurring during the period of the Weimar Republic. From the point of view of West German–Israeli relations this did not cause any difficulties since both parties now in the government were strongly sympathetic to Israel, rather more so, in fact, than the FDP. What was not known at the time was that the period of CDU-dominated government was coming to an end, to be replaced by a period of SPD-led governments. What was likely to effect a change in the long term was that the SPD under its leader Willy Brandt, now the Foreign Minister in the Grand Coalition, had adopted a programme for a revision of the Federal Republic's policy, which would end the bitter hostility between the latter and the Soviet Bloc and work for reconciliation between the two German states. It meant the abandonment of the policy of the reunification of Germany in the short term, which had been strongly opposed by the Soviet Union and the GDR, and it was hoped that détente would then come also to the relations between the Federal Republic and communist eastern Europe. The new approach, reluctantly agreed to by the CDU Chancellor Georg Kiesinger and his supporters in the Bundestag, became the official policy of the Grand Coalition but failed to achieve its aim. It will be argued in a later chapter that when the breakthrough did occur during Brandt's first SPD-led government that followed the Grand Coalition it contributed to a cooling of West German–Israeli relations.

Another phenomenon, however, manifested itself at the time of the Grand Coalition: the rise to prominence of two extremist groupings on either side of the political spectrum; they were the National Democratic Party (NPD) on the right already mentioned in this chapter, and on the left the so-called extra-parliamentary opposition. Both had openly anti-Semitic or anti-Zionist and strongly pro-Arab tendencies.

The NPD, an extreme right-wing movement, was neo-Nazi in all but name. That it called itself 'democratic' was dictated by the West German constitution under which parties which openly advocated the overthrow of the existing free democratic order could be banned by the Federal Constitutional Court. This had occurred on two occasions since the founding of the Federal Republic. The NPD clearly gained support as a result of the sudden economic recession which, as it brought memories of the 1930s, created a deep crisis of confidence. The party's rise to promi-nence – it succeeded in entering a number of state parliaments although no NPD deputies were ever elected to the Bundestag in Bonn – created misgivings not only in the Federal Republic but all over the world. People were asking whether the Nazis were returning to power. In actual fact the party's membership and even the votes cast during the Land elections in 1966–69 were insignificant compared to the support given to the two large parties now sharing the government in Bonn. True, Hitler's party also started from small beginnings in the 1920s, but the 1960s saw none of the difficulties of the 1920s and early 1930s such as national humiliation resulting from the lost war, financial ruin and mass unemployment. Because of the sensitivity of relations between the Federal Republic and Israel, however, the sudden rise of the NPD created an additional strain. Its importance was exaggerated by sections of the Israeli press, which prompted Pauls' criticism of 'bad reporting', and its very existence was exploited by politicians on the far right and left in Israel, who had always maintained that nothing had changed in Germany since Hitler, and that Israel should therefore have no relations at all with any Germans. There was ample evidence, however, that the vast majority of Germans were firmly opposed to a resurgence of Nazism. This was also accepted in Israel, certainly by the government. In the event, the economic recession was quickly overcome and the NPD dwindled to insignificance.

The 'extra-parliamentary opposition' (APO) was a spontaneous out-break of student discontent in 1967 manifesting itself in street demonstra-tions, sit-ins and in some cases riots. The student movement, largely

Marxist orientated – it was given the name of 'New Left' – was neither homogeneous in its philosophy nor centrally led, but inspired by various contemporary personalities such as Mao Tse-tung, Fidel Castro and his lieutenant turned dissident, Che Guevara. Its aim was to transform bourgeois democracy and bourgeois-dominated society but with only the vaguest concept of what the end product of such a revolution should be. The student revolt was by no means a purely German phenomenon; it was to be found in all western democracies and showed itself at its most virulent in France where it threatened for a brief period the stability of the government of de Gaulle. Its importance in the Federal Republic is due to the fact that together with the NPD it provided evidence of a new radicalisation in West German politics for the first time since the founding of the Federal Republic 20 years earlier, a phenomenon greatly feared because of the experience of the Weimar Republic. It derived its name 'extra-parliamentary opposition' from the idea – probably erroneous – that in the absence of an effective 'inner-parliamentary' opposition and because of the need for compromise among the two government parties which had no natural consensus, discontent had to vent itself outside parliament. The New Left supported liberation movements all over the world, including the North Vietnamese against South Vietnam and its American allies, the Iranian revolutionary movement against the Shah, and the Palestine Liberation Organisation. An important aspect of its ideology therefore was its anti-Zionism. It suppported the Arab cause and branded the State of Israel an imperialist aggressor. The NPD, like the Nazi party of Hitler's time, was fiercely anti-Semitic. By extension it took up the cause of anti-Zionism and gave strong moral support to the Arabs against Israel.[35] Both movements lost ground in the 1970s. The New Left brought about few real changes other than university reforms. It largely disappeared as a student movement, but not without helping to change the intellectual and political climate in the Federal Republic and elsewhere[36] and that in turn affected European and especially German attitudes to Israel. This will be discussed in the next chapter.

NOTES

1. Deligdisch, op. cit., p. 110.
2. The Times, 22 June 1965.
3. FAZ, 19 June 1965.
4. Deutschkron, op. cit., p. 347.
5. Vogel, op. cit., pp. 190–1.
6. FAZ, 20 June 1965.
7. Deutschkron, op. cit., p. 352.
8. FAZ, 9 Aug. 1965.

9. *Stuttgarter Zeitung*, 4 Aug. 1965.
10. *Neues Deutschland*, 25 April 1965.
11. *FAZ*, 15 May 1965.
12. *Bulletin* 1965/84, 14 May 1965.
13. Deutschkron, op. cit., p. 131.
14. Deutschkron, op, cit., pp. 129–30.
15. On 10 May 1965, see *Bulletin*, 11 Nov. 1965, p. 1448.
16. *Deutschland-Berichte*, vol. 1, Oct. 1965, p. 4.
17. Shinnar, op. cit., pp. 100–2.
18. Shinnar, op. cit., p. 106.
19. Hebrew: 'reparations'.
20. *Deutschland-Berichte*, vol. 1, Dec. 1965, p. 8.
21. *FAZ*, 22 June 1966.
22. *Ezekiel* 16:6.
23. Vogel, op. cit., p. 199.
24. Ibid (some commentators thought he was trying to placate opposition parties).
25. Vogel, op. cit., p. 199.
26. K. Adenauer: 'Bilanz einer Reise- Deutschlands Verhältnis zu Israel' in *Die Politische Meinung* vol. II, no. 105, June 1966, p. 15.
27. Vogel, op. cit., p. 200.
28. Vogel, op. cit., pp. 203–6.
29. *New York Times*, 7 July 1966.
30. *New York Times*, 7 July 1966.
31. *Neue Zürcher Zeitung*, 10 July 1966.
32. Interview with *Ha'aretz* on 12 July 1966 quoted by Vogel, op. cit., p. 207.
33. Interview with *Yediot Chadashot* on 22 July 1966, quoted by Vogel, p. 209.
34. K. Adenauer: 'Bilanz einer Reise-Deutschlands Verhältnis zu Israel', op. cit.
35. Anti-Zionism is not the same as anti-Semitism, being directed against the establishment of a Jewish national home or state in Palestine rather than against the Jews in general. But when because of the Nazi excesses against the Jews anti-Semitism became discredited after the Second World War and lost some of its efficacy as a political weapon, hardened anti-Semites often used the term anti-Zionist as as cover for their anti-Semitic views.
36. The intellectual upheaval of the late 1960s was called by some German scholars 'a kind of western cultural revolution'. See K. Sontheimer, *Die verunsicherte Republik*, Munich: Piper Verlag, 1979, p. 27.

9

The Six Day War and the Decline of the Relationship

Before continuing with the trends in West German foreign policy and their effect on relations with Israel an event must be noted which caused a great stir all over the world, but especially in the Federal Republic. In the early summer of 1967, while the Grand Coalition in Bonn was grappling with the economic recession and witnessing the first manifestations of the extra-parliamentary opposition, a new Middle Eastern crisis began, and on 5 June another war broke out between Israel and its Arab neighbours. Its importance for German–Israeli relations can be overrated if it is judged merely by the excitement and pro-Israel euphoria shown by the West German media and public. The Federal Government remained outwardly calm, declaring its neutrality in the conflict and reasserting its adherence to the principle, enunciated after the 1965 crisis, that no arms or military equipment may be sent from the Federal Republic to areas of tension. What was surprising, however, even if one considers the efforts of the Germans at reconciliation with Israel in the past, was the depth of public sympathy shown for the Israelis in this conflict, the concern for their survival and the admiration for their military achievements.

The complete story of why and how this conflict came about still remains to be told. It must of necessity be linked to the various currents in the Middle East and the world political system that have already been mentioned. Arab nationalism and President Nasser's ambition for the leadership in the Arabs' struggle for independence, intra-Arab relations, superpower rivalry and the state of play at that time of Israel's relations with its Arab neighbours must have all played major parts. What matters, however, for the impact of the war on West German–Israeli relations is the way the conflict was perceived by the public rather than the complexities of its background.

It may be expedient here to recall two items of the settlement of the previous Arab–Israeli war, the Suez conflict: that Egypt was obliged to allow free passage of shipping, including Israeli ships, through the Straits of Tiran which it controlled, so as to maintain the Israeli outlet to the Red

Sea and Indian Ocean through the port of Eilat; and that the Egyptian government had agreed to the stationing of a UN peace keeping force within Egyptian territory along the borders with Israel in Sinai and the Gaza Strip. One of the reasons for the latter was to prevent the incursion of Arab guerrillas into Israel.

In the middle of May the Egyptian leader began a fierce war of nerves against Israel. It started with the massing of Egyptian troops along the Israeli border in Sinai and the Gaza Strip and a demand to the UN Secretary-General, U Thant, to remove the UN peace keeping force from that area, a demand with which U Thant felt obliged to comply. After its departure, Egyptian forces occupied Sharm-esh-Sheikh on the Straits of Tiran and the following day Nasser announced that the straits would henceforth be closed to Israeli ships. On the same day the Israeli Prime Minister, Eshkol, demanded the restoration of freedom of shipping in the Red Sea and UN action to prevent the recurrence of fedayeen incursions into Israel. Whether by coincidence or as part of a concerted plan, guerrilla incursions into Israel had been increasing for some time on its borders with Syria and Jordan. On 28 May Nasser declared that the Suez Canal was now closed to all nations aiding Israel. All these events were accompanied by attempts at psychological warfare in the form of highly charged speeches by Nasser and the leader of the recently formed Palestine Liberation Organisation, Ahmed Shukeiry, then based in Cairo, threatening to destroy Israel and throw its Jewish inhabitants into the sea. Nasser declared *inter alia*: 'We will choose the time and the place to meet the confrontation with Israel' and referred to the struggle as a 'holy war'.[1] This propaganda, spread by Cairo radio and the Egyptian press, was vigorously supported not only by some other Arab states but also by the USSR whose government asserted that Israel was about to attack Syria and that it would have to reckon not only with the combined power of the Arab states but with the decisive resistance of the Soviet Union.[2] In those circumstances the Israeli government saw no other way than to prepare for war. On 5 June their air and land forces struck against the Egyptians in Sinai and the Gaza Strip. Syria immediately went into action against Israel in the north and Jordan, on the pretext that the UK and the USA had given military help to Israel, joined the two Arab states two days later.

The question of whether Nasser really wanted this war is still controversial. He himself stated that he would not fight unless the Israelis attacked first and claimed that he had given a promise to that effect to the UN Secretary-General when he demanded the departure of the peace keeping force.[3] He must moreover have known that Egypt did not possess the military capability to defeat Israel, even with the help of his Arab allies. He could not have counted on Soviet support other than what he already had: the supply of weapons and propaganda for the Arab cause. Financially Egypt was weak and it was still committed in the Yemen civil war. The US, although preoccupied with Vietnam and cool towards Israel after Suez, was unlikely to stand idly by if Israel was destroyed. Why then did Nasser go to the brink in provoking Israel? The view has been expressed that all he wanted was to inflict a diplomatic defeat and humiliation which would have greatly enhanced his standing in the Arab world. The final answer to this must be left to the judgment of history. What is certain is that the public and media in most western countries virtually ignored any arguments in his favour. To most of them he was at worst the aggressor, at best to blame for having provoked the war. Nowhere was this more strongly expressed than in the Federal Republic.

The Israeli reaction to the Egyptian provocation must be seen in the light of the country's short history and its strategic situation. Israel is a small country which had at that time a population of approximately 2.5 million. It had an unusually long frontier often through desert, which it could not possibly have guarded in its entirety. The distance between what was then the Jordanian frontier and the Mediterranean at its narrowest point just north of Tel Aviv was only nine miles and that point was close to the most densely populated and most important part of the country. There was no peace with Israel's neighbours, only an armistice punctuated perpetually by incursions of Arab guerrillas. The Arab neighbours did not recognise Israel's right to exist and some of their leaders were openly advocating its destruction. In those circumstances it is hard to imagine how Israel could have taken the risk of waiting to see whether the other side would attack or contemplated for long whether or not it had the right to pre-empt such an attack. This at least was the view expressed by the media and the public throughout the western world, even if the governments were more cautious.

Indeed, the major western powers declared their neutrality and this, to the surprise of many, included the USA, despite the vocal support given to the Arabs in Moscow and East Berlin. In the case of France, which had been a strong supporter of Israel since before Suez and had made an important contribution to Israeli defence, especially the air force, de Gaulle strongly disapproved of the Israelis being the first to strike. As became clear afterwards, however, this was the beginning of a general change of direction of French foreign policy in favour of the Arabs. President Johnson's inaction – although he did tell Egypt that its closure of the Straits of Tiran to Israeli shipping was illegal – was criticised in the western press as 'bungling'. It may well have been that, but the US government had been at pains to preserve its influence with the Arab world. Besides, all western states were concerned with the preservation of East–West détente and afraid that intervention on one side or the other in the conflict by either superpower or both could lead to an escalation or even a direct confrontation between the US and the Soviet Union.

This was certainly a preoccupation of the Federal Government as it had been with Adenauer at the time of Suez. Bonn waited nine days before cautiously expressing support for the freedom of shipping on the high seas after Nasser had closed the Straits of Tiran to Israeli ships. But despite all the West German government's insistence on reunification which had led to bad relations with the Soviet Bloc, none could have looked with equanimity at the prospect of a nuclear war between the superpowers, with the two German states in the front line. In any case the Germans were not ready to engage actively in world politics; they had not forgotten the debacle of two years earlier. As Foreign Minister Willy Brandt expressed it: the Federal Republic claimed the right to draw attention to the need to preserve peace, but would not intervene in regional conflicts.[4] Bonn was in fact still repairing the damage caused by the 1965 crisis when a number of Arab states broke off diplomatic relations with the Federal Republic. Negotiations with these countries about the re-establishment of diplomatic relations were still going on but had so far only succeeded with Jordan. There was some criticism at first in Israel of Bonn's reticence. But there was little the Federal Republic or any other west European medium power could have done in the conflict.

The main hope therefore lay in action by the superpowers to stop the war. Neither of them, it was believed, could afford its client state or states to be crushed. The Soviet Union had staked too much to want to see the Arab states go under or even be permanently harmed. For 12 years Moscow had been building up the armed forces of Syria and Egypt and pumped economic aid into them, besides giving strong moral support for the Arab cause against Israel. The diplomatic and propagandist support was now redoubled, giving the GDR government cause for bitter attacks on Bonn. It now claimed that Israel's aggression had been prepared with the help of the US and the Federal Republic,[5] and openly accused the Federal Government of having continued the arms deliveries to Israel after 1965 and having increased the shipments in March 1967, thus exacerbating the tension between Israel and its Arab neighbours. Chancellor Kiesinger reacted bitterly to what he called the campaign of slander against the Federal Republic and even sent a protest through its permanent observer to the UN General Assembly.[6] Despite the general deterioration in East–West relations brought about by the Middle Eastern conflict, however, the concern of both superpowers for the survival of their allies in the Middle East was seen right from the start of the war as a unifying factor between them that could lead to concerted action by them to stop the conflict.

When the fighting started the Chancellor reiterated his government's policy, which he referred to as non-intervention rather than neutrality. There may have been significance in the choice of that term because of the emotional upsurge of support and concern for Israel among the public and in the press: the word 'neutrality' would have implied indifference as to the outcome of the struggle.[7] Most politicians, especially among the ruling parties, the CDU and SPD, found it difficult to distance themselves totally from the wave of public sympathy for Israel. In a Bundestag debate following Kiesinger's statement the Foreign Minister, Brandt, put it this way:

> It matters to me very much to be able to emphasise as my personal conviction, with which I am not alone, that our non-intervention, i.e. neutrality in the sense of international law, cannot mean moral indifference or indolence of the heart.[8]

If that is as far as he as a member of the cabinet could go, another Social Democrat, Helmut Schmidt, then leader of the parliamentary SPD, went further:

> Much as we set store by the traditional friendship of our people with the Arab peoples, we must protest against the intention of their leaders to destroy Israel .... That nation has successfully built a democracy. We do not doubt its will to promote the peaceful development of its state and cannot listen to the threats to the existence of this nation without compassion.[9]

With this both leaders were reflecting the strong sympathy that the SPD had always shown for Israel. The leader of the parliamentary CDU, Rainer Barzel, after reminding the Bundestag that Israel was a state recognised by the UN, warned that there was an international agreement for the prevention of, and failing this, for the punishment of genocide.[10]

Despite the Federal Government's formal stance of neutrality the emotional atmosphere of the moment sometimes rubbed off not only on the politicians' statements but even on their decisions. Although the government had predictably continued its ban on the export of weapons to the Middle East and would not allow West German men liable for military service in the Bundeswehr to volunteer, as some had tried, to fight for Israel, it did allow volunteers to do civilian duties and for non-military goods to be sent to either side. They must have known that because of the attitude of the vast majority of the public this would greatly benefit the Israelis. Some of the goods at best were controversial. Thus the Israeli government had appealed before the fighting started for 20,000 gas masks for the civilian population. The request was justified by reports at the time that the Egyptians had used poisonous gas in Yemen. The plea led to a dispute within the West German cabinet with Kiesinger and Brandt being in favour, but Schröder, then Minister of Defence and his Secretary of State, Carstens, against. Schröder clearly considered them to be weapons, which therefore came not only under the West German rule but also under a NATO agreement not to export weapons to the Middle East.[11] The problem was solved by taking the gas masks from civilian rather than Bundeswehr stores, to which Schröder finally agreed.[12] But strong pressure from the press and public opinion may have convinced the Defence Minister that it was politically desirable to compromise. Whether it was deliberate or coincidental, the Israeli request for gas masks was a propaganda scoop, for the association of the possible gassing of Israeli civilians with the killing of millions of Jews by the Nazis in gas chambers was not lost on the German public. The gas masks were eventually sent to Israel but there was no occasion to use them and they were returned after the war. The sending of gas masks, like the dispatch of clinical or medical material, largely donated by private individuals or institutions could, of course, be justified on humanitarian grounds. This was less obvious where lorries were concerned. A 'rapid delivery' of these was requested by the Israelis after the outbreak of war. And it was explained by the Foreign Minister that lorries, like gas masks, were not on the government's list of weapons that may not be sent to areas of tension.[13]

If the position of the government could thus be described as benevolently neutral, the small opposition party, the FDP, was much stricter in its interpretation of neutrality. Its leaders protested against the dispatch of gas masks and lorries, which they regarded as war material that should not have been exported to crisis areas. They also called for a new declaration of West German neutrality by the government, stricter than those issued by the Chancellor and his Foreign Minister.[14] They were clearly sensitive to Arab protests about the sending of gas masks and lorries.[15] But then the FDP, though agreeing with the general direction of West German policy in the Middle East, never showed the same sympathy that the other two parties, the government and the public had extended to Israel. It will be remembered that its leaders were hesitant about the granting of large-scale restitution in the early 1950s, reluctant to support changes in the Statute of Limitations and opposed to the end to the establishment of diplomatic relations with Israel. This attitude never changed. There were, however, reasons for the reluctance of the FDP and of politicians of other parties, including Gerhard Schröder of the CDU and foreign minister under both Adenauer and Erhard, to give support to

Israel: the Arabs were becoming increasingly important not only as buyers of German exports but even more so as the main suppliers of oil to the industrial West. More will be said about this in the next chapter. Although it was by no means as clear-cut as that, there was to many people, and not only in the Federal Republic, a simple alternative, a choice between one side and the other: Israel, with its population of 2.5 million and its economy struggling for survival could not match the potential of the large, widely scattered Arab populations, their resources of oil and manpower, their potential markets and in the last resort military power which might one day put an end to the Jewish State. Politicians holding such views would justify them as political realism, considering that it was not in the national interest of the Federal Republic to pursue any policies which might offend the Arabs.

These were, however, the views of a minority. A large majority of the population in the Federal Republic showed by words and often by deeds their strong support for Israel.

That there was public sympathy in the Federal Republic at this time was not surprising. There had been public support before: over the question of diplomatic relations, of the West German scientists working for the Egyptian armaments industry, for example. Besides, public sympathy for Israel at the time of the Six Day War was spread all over the world, even in countries where governments were either lukewarm or pro-Arab.[16] It was concern for a small freedom-loving country, whose people had emerged from the ashes of genocide and destruction to build up their ancient homeland against heavy odds and who were now once again fighting for survival, heavily outnumbered, this time against their neighbours, who were regarded by many westerners as authoritarian and backward. But in Germany the emotional involvement with Israel during this brief but dramatic struggle was exceptional by its spontaneity and depth, affecting all sections of the population. One reason was that the Israeli people consisted to a large extent of the survivors of Hitler's extermination policies during the Second World War among whom were many whose native language was German. These were now seen to be in danger, for the second time, of extermination. The idea of Israel as a nation of survivors had, of course, been invoked before, especially in the case of the negotiations for the Luxembourg Restitution agreement and the dispute over the scientists working for the Egyptian government. But on this occasion the danger was dramatically displayed for all to see by the West German media. An appeal, initiated by Dr Adolf Arndt, an SPD member of the Bundestag, signed by many prominent citizens and published during the days of crisis preceding the outbreak of war set the tone. It read:

We cannot be silent when the Israeli people is threatened with genocide. The State of Israel is the last home of many who hail from our country and who escaped from the genocide committed against European Jewry and engineered by Germans. We urge all who bear public responsibility in our state, especially the Bundestag and Federal Government, not to stand silently aside but to throw in their weight for peace and stand by the Israeli people, morally and by peaceful means ...[17]

The full text was distributed by the German–Israel Society and appeared in most newspapers. It was a reaction to threats by some Arab politicians to destroy Israel. In its later paragraphs it also contained a guarded appeal, especially to young people, for volunteers and financial help for Israel. Some of these were already forthcoming: a week before the Arndt appeal there were reports of 'a wave of solidarity with Israel' in the Federal Republic. The Israeli consulate in Cologne was said to be 'assailed by volunteers'.[18]

Solidarity was shown also in other ways. On the day of the outbreak of war the Mayor of Berlin, Heinrich Albertz, declared that the sympathies of his city were with Israel, adding: 'Our thoughts are especially with those Israeli citizens who were driven from this city by the Nazis.'[19] The Berlin Senate donated DM100,000 for medicines.[20] On the same day it was announced in Frankfurt-am-Main that the city would donate DM30,000 of its budget to Israel.[21] Also, the West German Federation of Trade Unions (the DGB), as well as individual unions, decided to invest DM3 million in Israel bonds 'as a visible expression of solidarity with and confidence in Israel'. The DGB also appealed to members to participate in demonstrations by political parties, youth and church organisations for peace and the preservation of Israel, and to world opinion, the UN and all peace-loving nations to help bring about peace between Israel and its Arab neighbours.[22] There were innumerable other collections for and donations to Israel by private organisations and even individuals. There were demonstrations by students, political parties, trade unions and churches in Berlin, Frankfurt, Munich and other cities.

But there were signs that in all this concern for Israel there was something deeper than mere solidarity with the survivors of Nazi persecution, some emotions aroused by the Germans' repressed feelings concerning their past. During the post-war period much was spoken and written in public about the 'collective shame'[23] and responsibility of the German people as a whole for the Nazi atrocities during the war. Yet little was said about them in private. In very many families the subject was taboo, and children born after the war often complained that their parents and teachers were unwilling to discuss these matters. At the other end of the spectrum many positive aspects of German history such as great political and especially military achievements were, in the prevailing atmosphere of gloom and self-denigration, played down or suppressed in case they could reawaken the chauvinist and militarist tendencies that had been fanned by Nazism in the 1930s. The Six Day War, while not reviving these nationalist trends, briefly released emotions that had been repressed during the previous 20 years. Many Germans strongly identified with the Israelis during the period when Israel seemed in danger but especially during its victories. 'Our people, which had been suspected of self-satisfied ignorance and had suspected itself of succumbing to the egomania of affluence,' wrote *Die Welt*, in a leading article at the end of the war,

> gave evidence of spontaneous participation, signs of a vivid concern for the small courageous people of the Jews, signs of committed satisfaction about the successes, approaching miracles, of the Israeli stand. The advance of the Israeli troops was felt by most of the people with whom I spoke about the situation in the Near East as 'liberating' and 'heart stirring' as if it were the struggle of a brother people that had been won here .... A sultry murkiness of unclean sentiments had caused anxiety among us and among observers abroad. Into this oppressive atmosphere ... the war in the Near East struck like a purifying thunderstorm.[24]

By 'unclean sentiments' may have been meant the re-emergence of neo-Nazism in the form of the NPD; the 'purifying thunderstorm' could be the relief felt as a result of the Israeli successes after the tension caused by President Nasser's war of nerves against Israel; but more than that it was undoubtedly a release of tensions created by the Germans' unconquered past, emotions projected in the guise of a mix of rationalisations. Somehow the Israeli cause had become their cause, the victory their victory. Like the Germans in two world wars, the Israelis, the 'Prussians of the Middle East' had stood fast against overwhelming odds. Parallels with other events in contemporary history were invoked. Thus Moshe Dayan, the Israeli Minister of Defence and chief architect of the strategic plan of the war was hailed as the pupil of Field Marshal Rommel in desert warfare,[25] and as *Der Spiegel*, quoting *Bild am Sonntag* wrote: 'There are no more borders in the Berlin of the Near East', a reference to the reunification of Jerusalem under Israeli rule, and quoting from *Welt am Sonntag*: 'The Israelis did not get their freedom for nothing, neither will the Germans get theirs' – a reference to the struggle to free the East Germans from communist rule.[26] Even the language used recalls the virtually untranslatable metaphors of the communiqués of the Wehrmacht

154

during World War Two: shortly before the outbreak of the Six Day War *Die Welt* referred to the *Einsatzbereitschaft*[27] of the Israelis and at the end it was the *Neue Revue* which proclaimed that the 'desert foxes' had won, a name of honour hitherto reserved only for Erwin Rommel, Hitler's Commander-in-Chief in North Africa.[28] Finally *Blitzkrieg* and *Blitzsieg* were back in fashion. Clearly the Israelis had achieved the successes of which they, the Germans, had been deprived and they were impressed by these and, at second hand, perhaps even a little proud of them.

That the Jews, yesterday the enemies of the German people, were today the heroes of the 'brother nation' does not seem to have dampened the enthusiasm. On the contrary, a feeling of guilt, coupled perhaps with some relief that the truth about the Jews was out at last, increased the hero worship. 'Jews are not as anti-Semites wanted to see them. On the contrary, danger seems not to develop the evil but the noble qualities' according to *Der Spiegel*.[29] A leading article in *Die Welt* is rather less restrained: 'For 1900 years the most malicious lies have been spread about the Jewish people, that they were a people of cowardly sneaks keen only on trade and profit .... Now we have to revise 2,000 years of history ...'. One leader writer wrote that Israel had 'overcome our anti-Semitism'.[30]

There were newspapers in the Federal Republic which were more restrained and avoided what has sometimes been criticised as the 'adulation of the Israelis'. The *Frankfurter Allgemeine Zeitung*, while objectively reporting the appeals, donations, demonstrations and other expressions of support for Israel, concentrated more on an analysis of the international implications of the conflict. It was critical of suggestions that the war had solved anything, that 'a Gordian knot had been smashed which the politicians had tried in vain to loosen', or that it could be described as a 'purifying thunderstorm'.[31] Like *Die Zeit* it supported the neutrality of the Federal Republic from the beginning. As to the future, both papers advised the Arabs to accept the existence of Israel and the Israelis to give up the occupied territories.[32] 'The Israel army and people' wrote the *FAZ*, 'will have to dispense with triumph .... The Israeli army can only be seen as defensive .... It is accepted that gained territories can be bargaining counters but not war booty.'[33] *Die Zeit* went further and suggested a Jordanian free port in Haifa and an Arab corridor through the Negev in return for freedom of Israeli shipping to Eilat.[34] Both speak about solving the refugee problem and the possibility of some adjustment of frontiers. Both believed that superpower rivalry of the kind just experienced in the Middle East must cease as a prerequisite to lasting peace in the area.

While not all the press therefore indulged in lavish praise for Israel there were two small groups which – unlikely allies, it must be said –

actually supported the Arab side. One of them was the neo-Nazis, most of whom had joined the already mentioned ultra right-wing NPD and whose mouthpiece was the *Deutsche National-und Soldatenzeitung*, a strongly anti-Semitic paper which now wholeheartedly embraced the Arab cause to the extent of alleging Israeli war crimes and equating Israel, ironically, with the Nazis and Zionism with Hitlerism.[35] The other was the radical left-wing Socialist German Students' Federation, an organisation which had broken away from the much more moderate SPD and became prominent during the days of the extra-parliamentary opposition. This organisation had supported the cause of the Palestinians against Zionism and Israel already before the Six Day War. But though it was vocal and actively participated in street demonstrations, the majority of West German students were pro-Israel and openly showed their sympathies.

Although much of the enthusiasm for Israel reflected the emotions felt by the majority and the general atmosphere of the moment, some of the criticism is nevertheless justified. The emotionalism and lack of judgment are evidence of an immaturity of a kind that recalls the euphoria witnessed at the time of the takeover of power by Hitler in 1933. No comparison is intended here, the implications for the future being in 1967 infinitely more harmless compared with those of the events of 1933; only that in the two cases, both serious political events, the heart had taken over from the head in an extravagant way. The danger of an intoxication of this kind is that it may reverse itself in such a manner that the object of veneration becomes a target for contempt or even hatred. Such a reversal did not occur in Germany after 1967, but neither did the sympathy for Israel remain even at its pre-1967 level. In this war it was particularly apparent that the events revived memories of German history and evoked old self-images of the Germans. There is some truth in the statement by *Die Zeit* that the admiration of Israel's conduct of the war 'seemed to have revived sentiments of old German militarism',[36] and evidence that people saw in the Israeli military successes 'a drama of German history re-enacted and this time brought to a successful conclusion'.[37] There is also the feeling that the success of the survivors has in a way mitigated the horrors of the Nazi past: all's well that ends well. But these feelings focus on the German 'self', on the nation's recent experience and its attitude to that experience. They have, one is bound to conclude, less to do with an objective assessment of the virtues of Israel and its people. The admiration was thus unreal, spurious and therefore, not surprisingly, short-lived.

A word must now be said about the effects of the West German attitude on the Israelis. The very first Federal Government stance at the start of the war of strict neutrality was criticised by Israeli politicians and by the press, but this changed when news of the expressions of solidarity and of the sympathetic attitude of the media in the Federal Republic reached Israel. It has often been stressed here that the Israeli government for political reasons had accepted the Federal Republic as an ally if not as a friend, while large sections of the Israeli public remained hostile. But on this occasion the warmth of German public sympathy for Israel struck a chord with the Israeli public. The willingness to send gas masks and lorries was regarded as a sign that even the Federal Government was giving a little in its neutrality. The Israeli ambassador, Asher Ben-Natan, expressed himself favourably about the attitude of the West German media but above all was appreciative of the thousands of letters of sympathy and admiration addressed to him personally. The *Jerusalem Post* reported that the Middle Eastern crisis, in a single swoop, had done much to dispel fears of a Nazi resurgence in West Germany and had offered some reassurance of West Germany's future. It also quoted an Israeli Foreign Office source as stating that 'the impression gained from the press in recent weeks was that Holland and the German Federal Republic were the only two western European countries to have stood by Israel in deed as well as word during the present crisis'.[38] At a press conference on 22 June Shimon Peres, then on a special mission to the Federal Republic, said *inter alia* that in the recent conflict the Israeli public had learnt to esteem the West German attitude, both before and during the war. The feelings expressed in the press, on television and on the radio and the spontaneous reaction of the West German people had been a positive and important experience in the eyes of many Israelis.[39]

The Israelis responded to this with small but for them not insignificant gestures. At the end of 1967 the Israeli state radio broadcast midnight mass from Bethlehem in seven languages, including German. It was the first time that German had been used in an Israeli broadcast. Most of the cultural restrictions imposed as a result of a Knesset resolution in 1962 in the wake of the Eichmann trial were relaxed. There followed increases in youth exchanges and visits of teachers and academics from both sides;[40] these had so far been confined largely to visits by Germans since the Israelis still hesitated to allow their own citizens to visit the Federal Republic. Closer relations were also established between the Histadrut, the Israeli trade union federation, and its West German counterpart, the DGB.[41] And finally and perhaps most importantly, the Israeli press in general changed its tone towards the Federal Republic. The evening paper *Ma'ariv*, until then strongly anti-German, openly declared that it had reversed its attitude.[42] All this constitutes cultural or personal relations which do not necessarily affect the political climate at an official level. But as that had already become quite cordial the improvement at grass-roots level must be seen as helping to complete the normalisation process.

The sympathy for Israel, so strongly aroused in the West German public before and during the Six Day War, continued for a few months after peace was restored and then began to decline. Some of the underlying reasons belong to the realm of social psychology and are therefore beyond the competence of the political scientist or the specialist in international relations. But they must also be related to two phenomena which can be identified here: one, which was immediate, was disappointment with the outcome of the war, in that it had not brought the expected political solutions but had created some new problems. The other, more long-term, was that a new generation was growing up which was not involved in either the Second World War or its immediate aftermath and therefore held a different, more detached view of recent history. Both phenomena were not purely German ones, but were international.

That the Six Day War did not bring any noticeable alleviation to the problems of the Middle East was certainly disappointing to people all over the world, but with hindsight perhaps not surprising: the dramatic events of this very short war were calculated to raise excessive expectations. There had been hopes that the Middle Eastern war would end the uneasy armistice and lead to real peace between Israel and the Arabs, by which the chances of a future world war would also be reduced. That this did not happen was blamed by an increasing number of people in the democratic West on the refusal of the Israelis to surrender the territories they had occupied during the war.

Even while the fighting was still going on the world debated what should happen once the expected Israeli victory was achieved. Most serious newspapers wrote that Israel should withdraw from the occupied territories, some would have it do so unconditionally, as a gesture for peace, others were prepared to see Israel use the occupation of Arab land as a bargaining counter for recognition by the Arab states and for real peace. This latter approach was by far the most realistic but there were many difficulties. One of these was the unwillingness of the Israelis to give up East Jerusalem, the part that had been under Jordanian rule, because they regarded Jerusalem as the ancient capital of the biblical Jewish state and the focal point of the Jewish faith. They refused to see it divided again with the old city returning to Arab rule. But this was only one of the sticking points. An overriding reason for failure was the refusal of the Arab states to accept the existence of Israel. Another basic problem was the continued disagreement of the superpowers after they had briefly concerted their action to stop the war before there could be a total victory for Israel. For the defeat and humiliation of the Arabs was shared by one of the outsiders to the conflict, the Soviet Union, the superpower of the losing side. The Arabs could thus count after their defeat on the

USSR for the maximum of miltary and diplomatic aid the latter could give to restore them to international credibility as far as possible. Almost as soon as the war was over Israel, freshly home and dry from its military victory, was put in the dock at the UN by a coalition of Soviet Bloc and Third World countries, supported by some western states such as France, Israel's erstwhile ally, and Spain. Israel had to defend its pre-emptive strike against Egypt and was branded the aggressor by its opponents who demanded that it unconditionally relinquish all occupied Arab territory.

So the Gordian knot which the politicians had tried in vain to loosen was not smashed and, the war having failed to settle the Arab–Israeli conflict, the struggle reverted to the diplomatic arena. As time went on, however, the continued occupation created problems for the Israelis and these in turn drew world attention to Israel and the difficulties experienced by the Arab population in the occupied areas. An increasing number of Palestinians, most of them exiles from areas now either part of Israel or occupied by it, travelled to Europe as students, thus coming into contact for the first time in large numbers with Europeans, mainly young people. These therefore heard at first hand what many came to regard as the other side of the Israeli coin. Especially in the Federal Republic, where Israel had been for years uncritically admired, this created a shock which was soon reflected in the media. An Israeli academic wrote in 1972 after a period of research in the Federal Republic on West German attitudes to Israel: 'The German media began to criticise what was termed Israeli nationalism and chauvinism, the continued occupation of Arab territory, the annexation of East Jerusalem, the demolition of buildings, the curfew and the expulsion of Arabs to Jordan.'[43]

This links up with the second point, the more long-term question of the new generation to whom the Second World War and Hitler's extermination camps were history rather than experience. There could not have been such a relatively rapid change of attitude to Israel in the 1960s and 1970s if at that time there had not been – quite unrelated to events in the Middle East – changes in the political consciousness, especially among the young. The changes of attitude to Israel are said to have made themselves felt from 1968,[44] the peak year of the students' riots all over western Europe and of the extra-parliamentary opposition in the Federal Republic. What fired this revolt was that young people all over the West were no longer prepared to accept uncritically many of the long accepted views and values of their elders. In West German society where a degree of authoritarianism had survived despite the democratisation of the political system after the Second World War, the battle lines between the generations were particularly sharply drawn as those born after the war blamed their parents for Germany's Nazi past. Critical of their own society and political system, many young Germans felt sympathetic to peoples whom they saw as oppressed by autocratic regimes or suffering as a result of unjust wars or deprived of national self-determination. In their demonstrations they therefore turned against the Americans because of the Vietnam War and supported the revolutionaries against the Shah of Iran and the Palestine Liberation Organisation against Israel. Although the activists of this revolt, the New Left, were a small, if articulate and vocal minority, they could not fail to influence many others of their generation. Members of this post-war generation were now entering important areas of intellectual life, including the media, where they may even have affected the older workers already established there. These young people were not old enough to be responsible for what happened during the Hitler period and were therefore unwilling to accept such responsibility. As Medzini points out:

> Although the post-war generation of editors, all ardent anti-Nazis and Israeli partisans, still occupies the top positions, they are slowly being superseded by the 35–50 age group. Journalists in this group, generally Social Democrats or left-wing Catholics, although not radicals themselves, are nevertheless more sensitive to the views of their younger colleagues in the 25–35 age group who, in turn, are influenced by or expound the New Left doctrines which are intensely critical of Israel.

They believed that Israel 'should be more generous and humane in its attitude to the Arabs and more conciliatory in its efforts to promote peace in the Middle East'. The point is made that while until recently the West Germans had lavished their admiration on Israel's pioneering spirit, its successes in changing desert into agricultural land, in absorbing hundreds of thousands of immigrants in the face of threats by hostile neighbours, in the skill and valour of its soldiers in battle, they were now looking more critically at Israeli society, its internal stresses and its successes or failures to resolve them. They discovered that Israel as a nation was little different from other nations. Some took the line that Nazi crimes had no relevance to contemporary problems.[45] Why, then, have a special relationship?

These changes in the attitude of the public, however, had little effect on the government. Bonn at first did not move from its stance of benevolent neutrality, indeed it was said that some ministers privately expressed satisfaction with the outcome of the war. But the government now also saw as its task the re-establishment of good relations with the Arabs, a process that went back to the 1965 crisis. The fence-mending which started on the initiative of the Germans immediately after the crisis had not only been interrupted by the Six Day War but had actually suffered a set-back. Wounded pride on the part of the Arabs may have been one cause, the other was certainly the German public's categorically pro-Israel attitude and the Federal Government's one-sided interpretation of neutrality, seen by the Arab governments as favouring Israel. It was contrasted with the staunch moral support, described above, given to the Arab cause by the GDR government and the Soviet Bloc as a whole. The Federal Republic was therefore seen by the Arabs more than ever as a state allied to US imperialism, on the side of Israel against the Arabs, a 'western' state – an accusation now made openly but one which the Arabs had chosen to avoid in the past. The result was that despite great efforts on the part of the Federal Government, including offers of economic aid, little progress was made at first in wooing back those Arab states that had ruptured diplomatic relations.

On balance one may say that during the period just described the normalisation process in German–Israeli relations was completed. At government level this had already been achieved once the teething troubles connected with establishment of formal relations had been over-come. But the West Germans were now accepted with much more trust and sympathy by the Israeli public and media. This does not mean that all problems between the two nations had been cleared away. A degree of sensitivity on both sides remains and has led – and will no doubt lead in future – to occasional irritation, especially on the Israeli side. But as 'normal' relations between the nations go this is nothing unusual. The Israeli public had now caught up with the two governments and cultural and social exchanges developed more freely. In a sense the period of make-believe had ended or was about to end. There was less fear in Israel that the Germans might renege on their post-war stand or that they were full of unrepentant Nazis. West German economic support for Israel, the regular annual payments, continued, but attitudes became more critical. The West German euphoria over the Six Day War was the last irrational outpouring of enthusiasm over Israel of its kind, though the Federal Republic remained sympathetic and supportive. There was, similarly, an end to the illusion in respect of the Federal Republic and the Arab states: there never really was a special 'traditional' friendship between Germans and Arabs, only the usual friendship that springs from mutual need. This was evidently seen by both sides. Two years after the Six Day War, in the summer of 1969, a number of Arab states recognised the GDR, creating difficulties that were, however, no longer insuperable. But for this it will be necessary to return to developments in West German foreign policy in the next chapter.

NOTES

1. *FAZ*, 29 May 1967.
2. *Europa Archiv*, vol. 22, 1967, pt. 2 Documents, p. D300.
3. Kenneth Lewan: *Der Nahostkrieg in der Westdeutschen Presse*, Cologne: Pahl-Rugenstein, 1970, ch. III.
4. Willy Brandt's first statement on the crisis on 31 May, *FAZ*, 1 June 1967.
5. *Europa Archiv*, vol. 22, 1967, p. Z140.
6. Neither German state was then a member of the UN but they had 'observer' status.
7. The use of the word 'non-intervention' was criticised by the FDP opposition in the Bundestag.
8. Debate in Bundestag, 7 June 1967, *Bundestag Protokolle*, p. 5304.
9. Ibid., p. 5270.
10. Ibid., p. 5276.
11. *FAZ*, 2 June 1967.
12. *Daily Telegraph*, 2 June 1967.
13. *FAZ*, 8 June 1967.
14. *FAZ*, 7 June 1967.
15. According to the English Language *Egyptian Mail* (3 June 1967), the Deputy Secretary-General of the Arab League had protested that gas masks were war material.
16. These included France and even the GDR.
17. Deligdisch, p. 184.
18. *Le Monde*, 25 May 1967.
19. *Daily Telegraph*, 6 June 1967.
20. *Die Welt*, 7 June 1967.
21. *FAZ*, 6 June 1967.
22. *FAZ*, 7 June 1967.
23. Phrase coined by the Federal Republic's first President, Professor Theodor Heuss. The phrase 'collective guilt' was rejected, e.g. by Adenauer in his speech to the Bundestag on 27 Sept. 1951.
24. Mathias Walden in *Die Welt*, 10 June 1967.
25. *Rheinischer Merkur* quoted in *Der Spiegel*, 26 June 1967, p. 66.
26. *Der Spiegel*, 26 June 1967.
27. 'Readiness for action', *Die Welt*, 30 May 1967.
28. *Der Spiegel*, 26 June 1967.
29. Augstein: 'Long Live Israel' in *Der Spiegel*, 12 June 1967, p. 3.
30. *Bild-Zeitung* quoted in *Der Spiegel*, 26 June 1967.
31. *FAZ*, 10 June 1967.
32. *Die Zeit*, 9 June 1967. *FAZ*, 10 June 1967.
33. *FAZ*, 10 June 1967.
34. *Die Zeit*, 9 June 1967.
35. Deligdisch, p. 138.
36. *Die Zeit*, 7 July 1967.
37. Friedemann Büttner, 'German Perceptions of the Middle Eastern Conflict' in *Journal of Palestine Studies*, vol. 6, no. 2, Winter 1977, p. 66.
38. *Jerusalem Post*, 22 June 1967.
39. *Deutschland-Berichte*, July 1967.
40. Deligdisch, pp. 139–41.
41. Deligdisch, p. 142.
42. *Daily Telegraph*, 20 July 1967.
43. Meron Medzini: 'Israel's Changing Image in the German Mass Media', in *Wiener Library Bulletin*, vol. 26, nos. 3 & 4 (1972–73), new series nos. 28 & 29, p. 11. Article based on interviews by the author of journalists and academics.
44. Medzini, op. cit., pp. 8, 11.
45. Medzini, op. cit., p. 11.

10

Another Middle East War – enter the European Community

The changes in the relations between the Federal Republic and Israel in the late 1960s, as discussed in the last chapter, received a new impetus in the early 1970s partly as a result of *Ostpolitik*, partly because of other changes in the international political situation. The new West German policy regarding eastern Europe, initiated by Willy Brandt's SPD/FDP government, had little effect in itself; what caused a change was the further independence and self-assurance gained by the Federal Republic as a result of being able to shed the burden of the Cold War, which had persisted between it and the Soviet Bloc for several years after the global Cold War had diminished. For the first time since the founding of the Federal Republic in 1949 there existed a degree of amity and co-operation in West German relations with eastern Europe, resulting in a greater feeling of security. Not unnaturally, this greater self-assurance also rubbed off on the West German population. While there was continuing sympathy for Israel, there was also increasing impatience with the role of the eternal penitent whose atonement must continue indefinitely. The new, more self-confident attitude of the Federal Republic at a political level showed itself in its relations with all its western allies, especially the USA, where it led to disagreements. As far as Israel was concerned, the Brandt government, while still supporting the Israelis in matters which were essential to their well-being and security, was no longer prepared to allow itself to be hindered in its freedom of action in foreign policy to suit Israeli interests, as had been the case in the 1950s and 1960s. One reason why the Federal Government saw this as important was that it was trying to re-establish good relations with the Arab countries after a number of them had broken off diplomatic ties following the crisis of 1965. All this caused resentment in Israel where it was felt that the Germans were reneging on their moral debt to the Jewish State and people.

Quite independently of *Ostpolitik* changes were also occurring in western Europe. In Brussels applications by four countries, most notably the United Kingdom, to join the European Community had been on the table for several years. These applications had for long been vetoed by the French government under de Gaulle. With his departure from power in 1969 the way was gradually opened for these countries to be admitted and three of them joined at the beginning of 1973. The enlargement of the European Community from six to nine members quite naturally led to a change in the power structure within the organisation. France, for example, could no longer be certain of its dominant position once the UK had become a member, and the growing status of the Federal Republic was another factor. Both these countries had different foreign political interests from those of France. The differences might not have come out into the open had not the Council of Ministers of the European Community decided already in 1970 that the then six member states should try to establish a joint Community approach to certain international political questions. The French government, under the presidency of de Gaulle's successor Pompidou, who was continuing the pro-Arab policy first adopted by de Gaulle after the Six Day War, submitted to the member governments two proposals for a joint EC initiative to resolve the Arab–Israeli conflict. Both were distinctly pro-Arab and to the disadvantage of Israel, where they caused strong indignation. The more important proposal was made after the end of the October 1973 war between Israel and two of its Arab neighbours and at a time when some western states were threatened by an Arab oil boycott. The two French initiatives, which received some support within the Community, caused serious embarrassment to the Federal Republic. The West Germans, who had just toned down their relationship with Israel to win back the favour of those Arab states which had ruptured relations in 1965, were now put under pressure inside the EC to support declarations which went a good deal further towards an essentially pro-Arab policy than Bonn had intended to go. The EC policy declaration of November 1973 especially placed the Federal Government in a moral dilemma: either to dissociate itself from that declaration which was supported by most EC member states, some of which were threatened with losing the Arab oil supplies on which their economies depended; or to support the EC declaration and thereby give the impression that it was prepared to abandon Israel. The latter way would have had serious repercussions in the Federal Republic where, despite the more rational public attitude to Israel, it would have caused an outcry. In the event, neither EC initiative had any effect on the situation in the Middle East, where the only outside influence was that exerted by the superpowers. But at this time Bonn was walking the tightrope between its moral debt to Israel and its obligations to its pro-Arab western European neighbours, frightened by an Arab oil boycott. That situation was further complicated by the Federal Government's own particular brand of neutrality in the new

Middle Eastern war. The way it handled the conflicting pressures will form the main subject matter of this chapter.

Changes in the direction of West German foreign policy began in the early 1960s when, as already mentioned, détente between the superpowers replaced the Cold War of the 1940s and 1950s. Because the West German government at that time clung to its aim of political reunification through western strength against the two-states policy of the USSR[1] a relaxation of tension between the Federal Republic and the Soviet Bloc parallel with superpower détente could not develop. The result was that in the 1960s the Federal Republic's foreign policy was out of step with that of the USA, then its protector and closest ally. This led to occasional disagreements between Bonn and Washington and above all to doubts as to whether the military might of the US was still available to the Federal Republic as a bargaining counter to support Adenauer's policy of reunification through strength. As there was now little interest in the West generally in German reunification anyway, the Federal Republic was in danger of becoming politically isolated. That had one important effect: it forced its government to reappraise its attitude to the western allies and to take a more independent line. Its economic success in the 1960s also encouraged a new self-assurance. There was not, however, at that time a change in policy over the question of reunification; the strains between the Federal Republic and the USSR and Bonn's dependence on Washington therefore remained.

With hindsight it is easy to see that this state of affairs could not last long. There was no chance that the USSR, with its breakthrough in nuclear and rocket technology during the late 1950s and 1960s would succumb to western pressure for German reunification. Such pressure was in any case unlikely at a time when Washington and Moscow, having abandoned the Cold War because of the danger of a nuclear confrontation, were looking for areas of agreement, especially in the military field. There were many signals in the 1960s from eastern Europe indicating that Moscow would insist on the status quo in Europe: the building of the Berlin Wall in 1961 and the invasion of Czechoslovakia by Warsaw Pact forces in 1968 are examples. There was also increasing realisation in West German political circles that the old policy could not deliver reunification but that it was leading to greater alienation between the two German states. The SPD, under Willy Brandt's leadership, then still in opposition, was the first party to adopt a new policy of accommodation with the Soviet Bloc. The FDP soon followed suit. The CDU and CSU, however, maintained their opposition to such a change. That is why the Grand Coalition failed in its attempt to initiate *Ostpolitik*: CDU support was at first only lukewarm and was ultimately withdrawn.

Ostpolitik, when it was taken up seriously late in 1969 by the first left-of-centre coalition government since the war, with Brandt as Chancellor, was thus a belated, specifically German development of East–West détente. Unlike its predecessors the new Federal Government, with the CDU/CSU now in opposition, was prepared to accept the status quo in Europe as demanded by Moscow. That meant the continued division of Germany and the inviolability of existing frontiers, including Germany's eastern frontier along the Oder and Neisse rivers. Incorporated into the bilateral treaties which the Federal Republic concluded in the early 1970s with each Communist Bloc state were mutual assurances of non-violence in the settlement of all disagreements and of non-interference in the internal affairs of the other party. Reunification was thus consigned to history in the hope that the overcoming of the East–West conflict in Europe would one day make it possible. What mattered most to Bonn was that there would now be an official relationship with the GDR and that this in turn would lead to a better atmosphere and to talks on all matters of common interest, thus bringing about – so it was hoped – a reconciliation and gradual rapprochement as an essential preliminary to eventual reunification.

It would be over-optimistic to suggest that all the problems of division, which had haunted the Germans on both sides of the border for 20 years, could be eliminated as long as the ideological differences and the totally different political and economic systems remained. But the officially encouraged hatred and chicanery between the two states had created not only great personal suffering to many among the divided population, but was also a constant cause of irritation and political insecurity to both states. The new *modus vivendi* and the much improved relations removed a heavy psychological and political burden from the shoulders of the Federal Republic. It also brought considerable economic benefits since trade relations with the Soviet Bloc improved greatly. The Cold War which had remained between West Germany and its eastern neighbours was now over, so were its attendant dangers. The new situation gave to the Federal Government's foreign policy a much increased freedom of manoeuvre and with that greater independence and self-assurance. It is not too much to assume that this will have contributed to the changes of attitude of many West German citizens who felt that it was time the nation be freed from the burdens of its Nazi past.

The new enhancement of the international status of the Federal Republic was not greeted at first with universal enthusiasm in the West. The Americans at first welcomed *Ostpolitik* as a boost for détente. But détente did not mean the end of the East–West conflict or of the existence of the two big alliance systems. There was now concern in several western governments that the West Germans' new interests in eastern Europe might weaken their commitment to the Atlantic Alliance and to western Europe. When the treaties with the Soviet Union and the GDR had been signed the French government, despite assurances by Brandt, at first adopted a hostile attitude. President Pompidou expressed fear that the Federal Government was conducting a 'Rapallo' policy, a reference to an accord concluded between the USSR and Weimar Germany in 1922, in which it was suspected that Germany was trying to play off its eastern neighbours against its western ones. The new concern showed that in western Europe at least the fear of a powerful German state was not dead. There was no need for such a fear since the West German rapprochement with the Soviet Bloc was not and could not have been anything other than a normalisation of relations such as already existed between the communist and the other west European states. In no way was there a chance of a German–Russian alliance such as had existed between Imperial Germany and Tsarist Russia in Bismarck's days, nor was there a case for a Rapallo-type treaty. Quite apart from the Federal Republic's close economic and political involvement in the European Community it remained, and could not but remain, a member of the Atlantic Alliance and ultimately dependent on American protection.

The fear of a new 'Rapallo' soon subsided, but the suspicion that the Federal Republic might move towards a more neutralist position did not disappear entirely. By the early 1970s both German reunification and the unification of western Europe had been shelved indefinitely, confirming the Federal Republic as a permanent, independent state, at peace with its eastern neighbours and still committed to the West, even if with greater freedom of action. The Federal Republic had moved away from its position as the almost unquestioning client state of the United States, which it had assumed in the 1950s. The change became particularly notice-able when East–West relations sharply deteriorated again after the Soviet invasion of Afghanistan in the late 1970s. Bonn, fearing that it might lose the advantages it had gained through *Ostpolitik*, opposed the then militantly anti-Soviet stance of the Reagan administration, thereby arous-ing fears that the Federal Republic might take up some sort of 'neutral' position between the two alliances. Nothing like that occurred. What did happen was that the Federal Government no longer necessarily followed US policies and had in fact adopted a modest mediating role between the two superpowers. The Federal Republic, while remaining ultimately dependent on the power of the USA, was now strongly in favour of the preservation of East–West détente as it had a great deal to lose by a resharpening of the East–West conflict.

But this goes beyond the scope of this chapter and is intended only to illustrate the changes that occurred as a result of the Federal Republic's accommodation with eastern Europe and the doubts and fears these changes created. Israel did not remain unaffected by these either. There already was uncertainty about Brandt's attitude to Israel when he was still Foreign Minister in the Grand Coalition. He was reported on one occasion to have spoken of the need for a more self-confident representation of West Germany's national interests in the world, adding: 'I am not in favour of making a hair shirt into one's national costume.'[2] Such statements in themselves are not unreasonable and had been made before by political figures in the Federal Republic. But each time they created the suspicion in Israel that they presaged a fundamental change in West German attitudes towards the past and, by that token, changes of policy towards the Jewish State. Once Brandt was established as Chancellor of an SPD/ FDP coalition more fears were expressed in Israel that *Ostpolitik* could adversely affect relations with the Federal Republic. The reasons given were that Soviet anti-Zionism or pro-Arab policies might influence West German foreign policy towards closer relations with the Arabs at the expense of Israel, even that this might be the price Bonn would have to pay for better relations with the Communist Bloc; or that Brandt and his Foreign Minister, Walter Scheel, were returning to national power politics in which the German conscience over Israel had no place.[3] Finally, the Israelis were concerned that the foreign ministry in Bonn had been given to a leader of the FDP, a party whose attitude had been cool towards Israel from the outset.[4] That there was some shift in West German policy as far as Israel was concerned cannot be denied, but there was no need for Bonn to curry favour with the Soviets by embracing their anti-Zionist, pro-Arab policies; the new relationship between Bonn and Moscow was not of the kind where either side had to abandon its basic policies towards areas in other parts of the world. There was no possibility of a return to national power politics by the SPD/FDP coalition or any other West German government. For this the international political situation was totally unfavourable and this was repeatedly made clear by Brandt and members of his government. That a FDP foreign minister could affect to any extent the principles – though he might affect the style – of foreign policy is highly unlikely in a government based constitutionally on collective cabinet responsibility. It will be remembered that all West German governments but two had included FDP ministers and that not every CDU foreign minister had been pro-Israel.[5] What was correct was that the greater the independence and self-confidence of the West Germans and their government became, the more their conscience about the Nazi crimes was overridden by political concerns of the moment and the more they tended to regard Israel as 'a state like any other'.[6]

That there was this slow but noticeable change would not have been readily admitted by the West German Foreign Office; on the contrary, statements which implied that Israel would henceforth no longer be regarded as a 'special case' were usually followed by a denial that that was in the mind of the Federal Government and an assertion that the West German–Israeli relationship remained special in view of Germany's moral debt. This was a perhaps not very skilful attempt by the government to demonstrate that it wanted to pursue a 'balanced', that is to say impartial, policy in the Middle Eastern conflict without denying that conscience still played a part in its relations with Israel. It had, of course, happened in the past that the Federal Republic had put its own national interest before any consideration for Israel. The long hesitation in establishing diplomatic relations was one example. But the assertions that West German interests were paramount over those of Israel were increasing in frequency and soon began to be followed by actions which showed that the government meant what it said, while the explanations and statements about the 'special relationship' that followed appeared designed simply to soothe ruffled feelings. Looked at in that light there was little indication that such a special relationship existed any longer.

A more immediate reason for wanting to be seen to be pursuing a balanced Middle Eastern policy was the desire to re-establish formal relations with those Arab states which had broken them off in 1965. Strenuous diplomatic efforts were still being made by Bonn to this end in the early 1970s. Little progress had in fact been made when in the summer of 1969 Iraq, Sudan and Syria, followed within a few weeks by Egypt, recognised the GDR. It happened during the final months of the Grand Coalition which had made a first real, but unsuccessful, attempt at *Ostpolitik*. The Hallstein Doctrine was still formally part of West German foreign policy. How much the will to use this instrument had weakened was shown by the ambivalence of the West German reaction to the Arab states' decision. The Federal Government first reiterated the point made by previous governments that it must consider recognition of the GDR as an unfriendly act, but added that it would 'in such cases make its attitude and measures dependent on given cirumstances in line with the interests of the whole German people'.[7] Another statement read:

> The behaviour of the Iraqi and Sudanese governments in the German question has for the present destroyed the prospects for normalisation and improvement of relations between the Federal Republic and these countries ... The Federal Government will not be discouraged by the attitudes of the governments in Baghdad and Khartoum in its friendly sentiments vis à vis the Arab peoples but

will continue its efforts to cultivate and re-establish our good relationship with the Arab states.[8]

At about the same time Brandt, in the interview – already mentioned – with the Lebanese paper *Al Hayyat*, said that while the Arab states' action would not facilitate the normalisation of relations with the Federal Republic, nothing would prevent him from honestly striving for normalisation and the re-establishment of friendly relations with the Arab world.[9] As diplomatic relations with these countries did not then exist anyway there was little the Federal Republic could have done to 'punish' the offenders. But the expressions of anger and frustration and the threats were absent, replaced by the hope that relations would soon be resumed, suggesting that there was no longer any real desire to apply the Hallstein Doctrine.

It must be assumed, therefore, that many of the statements implying that Israel was 'a state like any other state' were made for Arab consumption, especially during the period up to 1972, when diplomatic relations between the Federal Republic and the Arab states that had ruptured them were finally restored. The Israeli government had on several occasions made it clear that it did not object to the Federal Republic entertaining good relations with the Arab states, even with those which were most militant in their hostility to Israel, provided of course that any friendship was not directed against or harmful to Israel and provided also that what the Israelis still saw as the special relationship with them was maintained. Over the latter point, therefore, there was some irritation. The Israelis continued to believe that the Federal Republic had a special duty to protect and support Israel where it could. This was especially possible in the Israelis' view in the diplomatic field. Frequent references by the Federal Government to Israeli withdrawal from the areas occupied during the Six Day War, for example, were not welcome even when they were linked to UN Resolution 242, passed in November 1967 and relating to Israeli withdrawal. That there were several interpretations to this resolution did not make matters easier.[10]

One such disagreement occurred in May 1971 when the foreign ministers of the European Community met in Brussels to decide on political co-operation by the Six in the international arena. It was the first such meeting, following an agreement to establish a 'consultation mechanism', concluded by the Six in Munich in November 1970, and the first serious attempt by the EC as a whole to venture into international politics. The aim was to achieve a Community standpoint or even Community policy among the six member states, hitherto known to have widely differing international interests. The meetings that followed led to a joint report, initiated by the French government and drafted by the political directors of the Community, setting out the Community's attitude to certain international problem areas, one of them the Middle East. Federal Foreign Minister Scheel saw in it 'a first manifestation of the will of the Six to participate throughout the world in the solution of the conflicts'.[11] It was never published, probably because it caused embarrassment to some member governments which could not agree with all of its content; but the gist of it became known.[12] It was said to include the following points: that Israel should withdraw from all territories occupied in 1967, except that some minimal border adjustments could be allowed; that demilitarised zones be created on both sides of the borders and these should be policed by UN troops under the control of the Security Council; that the old city of Jerusalem should be internationalised; that the refugee problem be regulated.[13] These points were diametrically opposed to the policy of Israel, whose government had made it clear after the end of the Six Day War that any Israeli withdrawal from the occupied territories could only be carried out in return for real peace and would have to be negotiated direct with the countries concerned. As for Jerusalem, it would remain under Israeli control, in fact, as Israel's capital. The Israeli government now expressed its anger, firstly that such a report, which seemed to favour the Arab point of view, should have been devised and secondly that a German government should apparently have put its imprimatur on it.

The content of the report, as far as it is known, must be seen in the context of relationships and policies prevailing at the time. The French government under de Gaulle had changed from a pro-Israel to a distinctly pro-Arab policy at the time of the Six Day War and had since supported demands made in the UN by the Arab states and the USSR for unconditional Israeli withdrawal from all occupied territories. Franco-German relations were strained as a result of *Ostpolitik* and French fears about the future of West German policy towards western Europe. In addition the European Community was negotiating about the accession of the UK and three other countries[14] to which France now reluctantly agreed after being urged to do so by the five other member states. The French government, which piloted the Middle East text and pressed for its adoption by the Six, may well have seen it as a way of reasserting French leadership in the Community in view of the impending accession of the UK, a major European power with which it had often disagreed, and also because of the prospects of an enhanced status for the Federal Republic. Taken together, these factors placed a heavy strain on Franco-German relations at the time when the EC report on the Middle East was being discussed by the foreign ministers. There were reports at one stage that relations had reached breaking point and that the French government had threatened to withdraw from all negotiations within the Community as it had done in 1965 because of disagreement with its EC partners. These reports, which may have been an exercise of inspired kite flying by the French government, were subsequently denied. But all in all there were enough reasons why the West Germans should have come under strong pressure on this occasion to follow the French line. Since the Council of Ministers, the main decision-making body of the Community, relies on a process of bargaining by the members of the individual governments in which each will press for its national advantage, it is possible that the Federal Foreign Minister, Scheel, substantially agreed to the French text in order to soothe French misgivings over West German policies in other, perhaps more important areas.

We do not know the final text agreed among all the participants, or whether it was substantially the same as the one originally proposed by the French government and severely criticised by Israel. But if it was it would be close to the interpretation of the UN Security Council Resolution 242 which requires Israel to withdraw from 'the' – meaning all – occupied territories. If Scheel had helped to ratify such a text, that could be seen as a departure from the originally declared West German stance in a way unfavourable to Israel. But from pronouncements and reports it is not certain that he did so.[15] According to a statement by the Israeli Foreign Ministry, the Israeli ambassador in Bonn, Ben Horin, received from Scheel a promise that he would resist the French proposals.[16] Herbert Wehner, leader of the parliamentary SPD, on a visit to Israel at the time, said the Federal Government would, if necessary, veto the French plan. Yet in Paris Scheel seemed to echo the French Foreign Minister, Maurice Schumann, saying that a plan had been agreed, that he had not yielded to Israeli pressure to veto it and that he had told this to the Israeli ambassador in Bonn.[17]

This whole confusing episode might be regarded as a storm in a teacup had not the Israeli government, worried that West German Middle East policy was veering towards the Franco-Soviet approach and that Israel was therefore about to lose an important ally in something that was of vital importance to its future, taken the matter very seriously. Israel, which claimed to know of the French-inspired proposals even before the foreign ministers entered into conclave to discuss them, had reacted promptly and sharply. As early as 1 May the Israeli Foreign Minister, Abba Eban, told Wehner, then visiting Israel, that he hoped that the European Community's Middle East document would not follow what he called 'the French adoption of the Soviet-Arab line of a return to the 1949 borders'. The government sent notes to the Five (the members of the EC except France) warning them that any attempt to dictate specific terms to the parties of the Middle Eastern conflict was calculated only to lessen the chances of a genuine peace agreement and recommending that they should adhere to the principles of UN Resolution 242, which in the Israeli interpretation meant 'withdrawal from territories' occupied by Israel. Any other proposals, he hoped, would be vetoed by Israel's friends.[18] The foreign ministers' document, it was claimed, went well beyond the UN resolution. When the foreign ministers' conference was over and it was suspected that Scheel had yielded to the French government, Eban took the unusual step of expressing his anger to the West German ambassador-designate, Jesco von Puttkamer, as he was presenting his credentials, at Scheel for having allegedly reneged on his promise to Ben Horin that he would veto the French proposals.[19] The *Jerusalem Post*, in a leading article, commented that Israel expected its friends in Bonn, Rome, Brussels, The Hague and Luxembourg to understand that the real motive behind the French initiatives was not that of a disinterested party, since France was 'unfortunately deeply involved with the Arabs'.[20] A week later the same paper wrote that 'behind the German action there was evidently political horse-trading'.[21] That last comment may well be true.

The writing of the EC report of 1971 is of no real importance in itself and there was no sequel to it, neither did it lead to any concerted action by the European Community in the Middle East. Its significance lies in the way it came about and the manner in which it was handled, especially by the Federal Government. If a show of community solidarity is achieved simply as a result of a quid pro quo by one member to another or simply to soothe the other government's anger or suspicion, then such a show of solidarity is unlikely to enhance the influence of a body like the EC in world affairs, as it highlights the divergence of interests among its members rather than agreement, let alone unity. So far as the 1971 EC report is concerned, all indications point to this having been the case. With regard to the relations between Bonn and Jerusalem the episode indicates 'normalisation' in the sense that, at least from the West German point of view, a special relationship no longer existed. As *Der Spiegel* expressed it at the time:

> Normalisation for Bonn today means that it can permit itself open conflicts of interest with the Israelis and that the responsibility for the murders of the past, though this is not denied or minimised, must no longer restrict the freedom of action of the German government in the international political field, at least not as a matter of principle.[22]

The contretemps over the EC paper subsided when the Federal Foreign Minister Walter Scheel visited Israel in the summer of 1971. He

173

played down the importance of the paper and assured the Israeli government that *Ostpolitik* would not impair West German freedom of action, that Bonn would not deviate from the UN Resolution 242 and that peace in the Middle East must come about by freely agreed negotiations.[23] A new rift developed, however, when a year later 11 Israeli athletes were killed by Palestinian terrorists at the Munich Olympic Games.[24] The massacre was a particularly severe blow to West Germans, not just because Israelis were involved but because the 1972 Olympiad re-awakened memories of the 1936 games held in Berlin, which had been regarded as an exercise in Hitler propaganda and during which there had been evidence of racial discrimination against coloured and Jewish sportsmen and women. Many Germans regarded this first Olympiad on German soil since 1936 as an act of redemption. The tragedy of September 1972 therefore created deep feelings of sorrow and a certain sense of failure in the Federal Republic. The acrimony between the two countries that followed arose over the question of security. But while the Israeli press had expressed anger with the German authorities for not adequately protecting the athletes, the Germans earned the respect of the Jerusalem government for refusing to give in to the terrorists' demands, an approach which accorded with the Israelis' own method of dealing with terrorists. What caused greater dismay in Israel, however, was the action of the federal authorities two months later when the three surviving terrorists, who had been captured at the time of the attack, were exchanged for 20 civilian hostages taken when a Lufthansa plane was hijacked by other Palestinian terrorists. The kidnapping of the plane was precisely intended to free the three Palestinian survivors of the Munich attack and the kidnappers' success was quickly noted by the rest of the world. This time government and public in Israel were united in condemnation of the Germans' action. The Prime Minister, Golda Meir, called it 'an affront to the human intelligence' and Yigal Allon, the Deputy Prime Minister, 'a disgraceful act bordering on cowardice and unworthy of enlightened civilised society'.[25] Demonstrating Israeli students carried a placard reading: 'Hitler murdered Jews, Brandt protects Jew murderers'.[26] *Ma'ariv* wrote: 'Not Arab courage but German cowardice assured the desired result'.[27] *Der Spiegel* commented: 'To acknowledge the obligation to exchange the prisoners for hostages was for Jerusalem tantamount to recognising the power of the guerrillas', and, in answer to Israeli accusations of cowardice: 'This ignores that Israel was the first state to exchange Palestinian terrorists for planes and passengers' (at Algiers in 1968 and Damascus in 1969) and 'that the last attempt to avoid the exhange of hostages led to the death of nine Israeli athletes'.[28] International terrorism of this kind, for the purpose of freeing prisoners, or for other ends, has of course become a fact of life since the early 1970s. Different governments have dealt with such situations in different ways, but in a majority of cases resistance by the authorities to a threat by the terrorists to kill the hostages unless their demands are fulfilled has ultimately led to the surrender of the terrorists and the sparing of the hostages. But this is never certain. The Munich massacre of 1972 was one of the first of its kind. The Germans, acting with little experience except their own failure to save the Israeli athletes, decided not to risk the lives of 20 innocent civilians taken captive by the terrorists. In the light of this, the Israeli reproaches, especially the references to Hitlerism, seem unjust. They caused anger and dismay in the Federal Republic and were not calculated to improve the attitude of West Germans to Israel at a time when, while the Israelis were still counting on German support for moral reasons, an increasing number of Germans in the Federal Republic had already become impatient with them.[29]

Meanwhile West German policy regarding Israel and the Arabs continued to be 'even handed'.[30] This was now the official German attitude towards the Arab–Israeli conflict and the Middle East in general. As the Federal Government had no power and little influence over Middle Eastern affairs, the 'evenhandedness' was confined largely to moderate statements telling each side the minimum of what it wanted to hear: for the Israelis it was that they were entitled to live in peace within secure borders and that the German–Israeli relationship remained affected by what Germans had inflicted on the Jews in the past; to the Arab states it was a reassertion of the 'traditional German–Arab friendship' and occasional cautious references to the Arab refugee problem and the rights of the Palestinians. Besides, both Brandt and Scheel reiterated statements they had made in defence of the European Community initiative in 1971 to the effect that because of the proximity of the Middle East to Europe, peace in the Middle East was of great importance to all west European countries. To this Scheel added the argument that since membership of the EC had brought the Federal Republic closer to the Mediterranenan, it had a great interest in the events of the Middle East. By and large all these statements were acceptable to both sides, though one cannot help feeling that they must have created some doubts in the minds of public figures in Israel, where a much more positive attitude to their country by the West German government was still expected.

In June 1973 Chancellor Willy Brandt paid an official five-day visit to Israel, the first ever by a West German Chancellor in office. It was no coincidence that Scheel during those days visited a number of Arab capitals. Brandt's reputation was such that despite the ill-feeling caused seven months earlier by the release of the Palestinian terrorists responsible for the Munich massacre he was welcomed with warmth and affection.[31]

He was received in the Knesset on the day of his arrival and was later awarded an honorary doctorate for his efforts for peace by the Weizmann Institute, the most prestigious academic and research institution in Israel.

The Prime Minister, Golda Meir, had lavish praise for her visitor, acknowledging his courage and conviction in treading new paths

> in order to bring East and West closer together and to persuade them to live together in the one world we have ... Perhaps the victory you have won is the greatest that can be granted to man ...[32] I am sure that your people are proud of you ... proud also of the place you have won in the world of today.[33]

This was a generous tribute to a German public figure, more generous than one was accustomed to from the polite but correct statements by Israeli statesmen in the past. In his reply Brandt made a point that had not been made in that form before; referring to Germany's treatment of the Jews during the Third Reich he said:

> The recognition of responsibility for the crimes for which Nazism abused the name of Germany was for us the decisive act of inner liberation, without which our external freedom would not be credible or reliable ... The three decades which have passed since the horrors have not allowed us to forget what must not be forgotten.[34]

The reference to 'inner liberation' indicates that Brandt's thoughts were those of Adenauer, Erhard and many others, even if they were not expressed in those words. Taking responsibility and acting accordingly were dictated by a moral need, a need to redeem the nation in its own eyes, so that it could survive as a self-respecting as well as respected society, and not simply by the desire to fulfil the exigencies of a foreign policy or the national interest.

In more practical terms, the Chancellor noted that co-operation between his country and Israel had developed well in many fields and that it should be further extended. He reiterated that his government did not see itself called upon to take sides in the Arab–Israeli conflict or to adopt the role of a mediator. The German interest was clear: it was the achievement of a peaceful solution, negotiated by and acceptable to the parties directly involved. So far as the EC was concerned, he said he would 'bear Israel's economic and political interests in mind when the newly extended Community sets about its task of comprehensively and objectively determining its Mediterranean policies'. He repeated that West German–Israeli relations 'must be seen against the sombre background of the Nazi terror. That is what we mean when we say that our normal relations are of a special character'.[35]

The speeches and comments show every sign that the visit took place in a friendly and relaxed atmosphere and that there were no major disagreements. Indeed nothing Brandt said or did could have given offence to any Israeli government. He did not repeat previous statements of his that one cannot continue indefinitely to atone for one's sins, but neither did he spell out in what way West German–Israeli relations might be further extended. Nor did he say, apart from a reference to his government not wishing to take sides in the Middle East conflict, what form its objectivity or evenhandedness would take. There was, it appears, no response to an often expressed Israeli view that the Federal Republic should continue to support the Jewish State in more concrete terms, especially so far as its economy and security were concerned. Golda Meir concluded her speech at a dinner given by Brandt in her honour with a veiled criticism: 'Sometimes it looks as if it were fair to treat everyone equally. But we know that abstract and legalistic equality is not always a mark of fairness.'[36] Brandt's reference to 'normal relations of a special character' was not lost on the Israeli public. There was criticism in the Israeli press. The *Jerusalem Post* wrote: 'Israel is of the opinion that the special relationship must continue for many years, based on Germany's moral debt. This should be shown by a generally benevolent attitude by the Germans whenever the need or opportunity arises.' As examples it listed the UN, the EC, the plight of Russian Jewry and economic aid.[37] The independent *Ha'aretz* wrote: 'The special relationship should extend beyond evenhanded relations, should be different from those with other states and should include the welfare and security of Israel.'[38] In Germany the *Frankfurter Allgemeine Zeitung* acknowledged that the Israelis had doubts about the Brandt visit. The Israelis, it wrote, were apparently interested in votes in the UN, but no promises of pro-Israel votes were made.[39] It quoted the Israeli evening paper *Ma'ariv* which said:

> All interpretations offered so far of Brandt's statements and all reassuring words of diplomats cannot detract from the basic fact that behind the gestures and smiles a West German policy is developing which undertakes no responsibilities for the terrible crimes which have been committed by Germans against us Jews. The new scheme of relations which the Chancellor tried to establish during his visit has only one significance: to free Germany of all political obligations that result from the Nazi inferno.[40]

This seems at first sight an excessively pessimistic assessment of the Brandt visit, a reaction not uncommon in the Israeli press. But then the sayings at official visits rarely consist of other than pleasantries, while more significance is attached to what was not said. The Brandt visit, as far as this book is concerned, is important less for what it was or for what was said than for what followed later that year.

The new Middle East War broke out on 6 October 1973. On this occasion a surprise attack was launched by Egypt and Syria against Israel on the Suez Canal in the south and on the Golan Heights in the north.[41] Because it was the two Arab states which had the benefit of surprise, the Israeli secret service having failed to detect their opponents' preparations, Israel was not at first ready to meet the onslaught. This meant that the Egyptians were able to cross the Suez Canal and establish bridgeheads on the Sinai side and the Israeli forces were therefore at first under considerable strain. As in the case of the 1967 war both sides had superpower support, but as on this occasion material losses on both sides were heavy and the war lasted longer, both sides had to be substantially resupplied with weapons – the Arabs by the USSR almost from the start, the Israelis by the US during the latter part of the war. For geographical reasons the Americans found it necessary to supply Israel from NATO stocks in Europe.

The international scenario was at first similar to what it had been during the Six Day War. The Soviet Bloc was again supporting the Arabs with propaganda as well as arms, including the GDR which again condemned Israel as the aggressor and the Federal Republic as Israel's accomplice: this despite the recent signing of a treaty between the two German states in which they undertook to improve their relations and gave each other de facto recognition for the first time since the Second World War.[42] The US, though siding with Israel, at first tried to give an appearance of neutrality in the hope of effecting a ceasefire through negotiations with both sides and with Moscow. In West Germany Willy Brandt declared that the Federal Republic would be neutral but added that all countries of the Middle East must be guaranteed the right to exist.[43] The European Community, in a declaration, for the time being contented itself with a plea to both parties of the conflict to end the war.[44] But very soon a new factor entered into this scenario: the threat of an Arab oil boycott against those states which supported Israel. Since Arab countries supply most of western Europe's oil needs and industry depends on oil, this threat instilled fear into west European governments and into the populations as well.

The oil boycott, combined with a 5 per cent cut in oil production, was decided by the ministers of the member states of OPEC on 17 October in Kuwait. It singled out two countries, the US and the Netherlands,[45] for a complete cut in oil supplies from Arab states in the Middle East and threatened others, including the Federal Republic, with similar treatment if they supported Israel in any way. The threat, which in view of the dependence of industry on oil had to be taken seriously by the affected countries, was intended by the Arab states to force the West, especially western Europe, to support the Arab cause against Israel. That this was possible was an indication that Arab solidarity had reached a stage where the Arabs could seriously challenge western economic and political supremacy in a region that was of extreme importance to the developed world. The more Egypt and Syria found themselves militarily in difficulties, the greater the chance that the 'oil weapon' would be vigorously applied. The one country that had least to fear from the boycott was the US because they not only had their own resources but could also continue to draw on others, especially in the western hemisphere. Only a relatively small proportion of their oil came from Arab countries. They were therefore in the best position to ignore the boycott. During the latter part of the war they started to supply the Israeli forces with weapons in order to save Israel from possible defeat. This brought them into conflict with western Europe because of the need to use European territory and installations, and especially with the Federal Republic when it was discovered that arms from American NATO stores in West Germany were openly being loaded on to Israeli ships in a West German port.

The circumstances surrounding this episode were complex. Behind the German reaction lay the determination of the Bonn government to adopt a strictly neutral position. This was not new: it had been so at the start of the Six Day War, but then neutrality soon mellowed to 'non-intervention' and what ultimately emerged was a neutrality very benevolent towards Israel. This time the Federal Government wanted to be seen to be taking neutrality seriously because the situation had changed to the extent that Bonn had now re-established formal relations with all Arab states and there was a threat of an oil boycott. When the American arms deliveries started in mid-October Scheel reminded the US ambassador in Bonn, Martin Hillenbrand, of West German neutrality and was apparently assured by him that the Americans would respect it.[46] Following on this, the Federal Government through its ambassador in Cairo told the Egyptian Foreign Minister, Ismail Fahmy, that no more US arms would reach Israel from the Federal Republic. This assurance was given despite the fact that American arms from NATO stores were being flown to Israel from US air bases in the Federal Republic. The airlift, by which by far the largest quantities of American war material were sent to Israel, and not only from West Germany, had been kept secret between the United States and the countries from which the arms were sent. The Federal Government, it was assumed, was being kept informed. The day after the assurance had been given to Cairo, however, the news broke that two Israeli freighters had been loaded with tanks, lorries and guns at Bremerhaven, the transatlantic terminal of the US military forces based in Germany, in full view of the public and press photographers, one of whom had been arrested by an American military policeman. The two ships had flown Israeli flags though their names had been painted out. Of this the Federal Government claimed that it had been unaware until the appearance of press reports.[47] Stung into action by this embarrassing news, it now protested vigorously to the US ambassador and his chargé d'affaires, Frank Cash, telling them that the strict West German neutrality in the Middle Eastern conflict required the government to debar arms deliveries from US depots to any of the belligerents if they involved the territory or facilities of the Federal Republic, adding later that 'the Federal Government assumed that American arms deliveries from and over the Federal Republic had now finally ceased'.[48] Both statements seemed to suggest that the ban applied not only to ships but also to the airlift. But a day later the Federal Government spokesman, Dr Armin Grünewald admitted at a press conference that Bonn had no power to control US aircraft taking weapons from West Germany to Israel.[49] It seemed the West German government now argued that American air bases in Germany were regarded as extra-territorial, meaning that the Germans had no jurisdiction over them. The way seemed open to continue the airlift of arms from the Federal Republic to Israel as long as was necessary for the Americans to replenish the Israeli arsenals depleted by the war.

The anger of the Federal Government was aroused not so much by the fact that arms were being sent to Israel from West Germany but by the manner in which it was done and the embarrassment it caused. Despite the American assurance to the contrary, the US, so it seemed, were openly, indeed publicly flouting West German sovereignty at the worst possible moment, with the news of the Bremerhaven loading following hard on the assurances given to Cairo. Bonn officials spoke of 'extraordinary American tactlessness'.[50] But once again the Germans had been caught on the horns of their dilemma over their conscience about Israel and their desire to establish good relations with the Arab states. In order to put the whole episode into perspective it is important to note the date of the protest: it was 25 October, two days after the ceasefire was due to have taken effect and on the day when it actually did begin to hold.[51] It seems certain that the Germans, though they had known about the airlift since the middle of the month, allowed it to continue and would have been content to let it go on without protest had it not been for the embarrassment of the Bremerhaven incident. But by then the fighting was over. At the beginning of the war Brandt had said:

> I will not hide from anyone how deeply this tragedy hurts us Germans ... because we must remind ourselves that here [i.e. in Israel] the lives of a surviving remnant are menaced, lives of people whose families were swallowed up in the great Holocaust. Their right of existence and that of their people is involved.[52]

The same day, it was reported, he exclaimed at a cabinet meeting: 'Woe, if Israel is knocked to the ground! What effect this would have on our people!'[53] The German conscience was not ready to bear a new reproach, that of trying to prevent essential war material from reaching Israel at a moment when the Israelis were fighting for their lives.

Unlike the government, the West German public did not have to agonise over a political dilemma. Nevertheless, although the people by and large again rallied to the Israeli cause, the enthusiasm of 1967 was absent on this occasion. There were some expressions of support. The mayor of Berlin, Klaus Schütz, participated in a demonstration of solidarity with Israel,[54] and the German–Israel Group of about 100 Bundestag deputies passed a resolution pledging solidarity, which they sent to the Knesset. It asserted that Israel had the right to exist within secure borders and by implication blamed the Arabs and the USSR for aggression.[55] Opinion polls also again showed a majority for Israel: the Allensbach Institute now recorded that 57 per cent of the population favoured Israel[56] while six months before the outbreak of the war the figure had stood at 37 per cent. But these polls do not reflect the expression of feeling such as was shown in 1967. It was much more subdued in 1973, both in the media and among the public. Yet there was concern about Israel's survival. This led to criticism about the Federal Government's interpretation of neutrality, especially from the CDU/CSU opposition. Die Welt wrote: 'Military neutrality in the Middle East there must be. But Israel cannot for us be a state like any other. Moral neutrality would become immoral cynicism.'[57] The Deputy Chairman of the CSU parliamentary party, Friedrich Zimmermann, reproached the government for having by the stoppage of American arms to Israel departed from the road of neutrality and committted an unfriendly act against Israel.[58] But all this was nothing like the irrational euphoria of the previous war. The more muted response by the public and media on this occasion may in part have reflected the more sober assessment of Israel and the Middle East by many sections of the people, which had developed since 1968. In the last chapter the point was made that much of the excitement in 1967 was due to the unexpectedly brilliant military successes of the Israeli forces, with which the West Germans somehow identified. Although there were Israeli successes of no mean calibre in October 1973, they came only in the shadow of early reverses and even then took longer to materialise as the Israelis had to overcome stiff Egyptian resistance. The Bonn correspondent of the *Jerusalem Post* expressed it thus:

> The change from the unparalleled enthusiasm about Israel during the Six Day War of 1967 was all too obvious. In part this was the result of the unexpected Arab advances and combat spirit in the new fighting. It was also the realisation that the war could drag on, possibly involving the big powers. And underlying it all was clear disappointment within the government and press alike over the lack of progress in past years towards lasting peace in the Middle East. This led a number of major papers to conclude that Israel itself, through its tough stand, had driven the Arabs finally to attack.[59]

That seemed a fair assessment, though one should add to that the fear, at least among the public, of an Arab oil boycott such as had already been imposed on the Netherlands.

An unpleasant by-product of the Middle Eastern war was the major row between Bonn and Washington caused by the Federal Government's ban on American arms for Israel and the protests made to the US over the Bremerhaven affair. These protests were indeed strongly worded and the way they were made was untypical of the method normally adopted even over disagreements by these two countries which had always been close allies. They seem to have been effective in one sense, for the third Israeli ship calling at Bremerhaven for arms had to leave empty. Ironically the other west European countries had been even stricter in their application of neutrality, if that is what it can be called. For the UK had banned weapons for Israel from its territory right from the start of the war while allowing arms to go to Arab states. France, Italy, Spain and Greece had similarly refused the transit of arms bound for Israel via their territories.[60] This of course highlighted the fundamental divergence of policies between the US, determined to safeguard the continued existence of Israel and able to resist the Arab boycott, and most west European states which were either, like France, already committed to a pro-Arab policy, or afraid because of the threat to cut their oil supplies. But the events also once again raised the question of the role of the senior partner in NATO: to what extent could the US use NATO facilities and military installations on their allies' territories in order to pursue what is seen as their own national interest, especially in regard to countries that lie outside the NATO area; and, generally, could they use their allies in whatever way without adequate consultation? This last point caused particular dismay in Europe which was not allayed by the US action a few days after the

Bremerhaven incident of issuing a NATO alert to all US forces in Europe without seeking the views of the European allies.[61]

The handling of the crisis by the Federal Government was less than adroit. Its main mistake was that it over-reacted at the news that Israeli ships were being loaded with weapons for Israel at Bremerhaven. It is easy to see that the Federal Government could have made its point to the US more discreetly through diplomatic channels, in which case it would have created less publicity and therefore less ill-feeling in both the US and Israel.

As in the case of the 1965 crisis, the West Germans drew upon themselves criticisms from all sides: from Washington both because of the substance and the vehemence of the protests, from the Arab states for having allowed NATO arms to leave German territory for Israel and from Israel for having tried to stop them leaving. But it is also fair to say that the Americans made life difficult for the Federal Government by not warning it of the imminent arrival of Israeli ships and by loading them without the precaution of secrecy, if indeed it was necessary to send the weapons by this method. The Germans, embarrassed by the impact that the loading at Bremerhaven would make in the Arab capitals, feared not so much for their oil as for the development of German–Arab relations, which had only recently been re-established, after long and painstaking diplomatic efforts. The American reaction seems to indicate that on this occasion Washington unnecessarily took its European partners for granted.

The US government indeed made no secret of its displeasure over the Europeans' attitude. An early reaction was that Congress might demand a cut-back of US forces in Europe. A Pentagon official said bitterly: 'We feel there was not one country in Western Europe that, if pushed to the wall, wouldn't have let Israel go under.'[62] But anger in Washington was expressed most strongly and openly against the Bonn government even though the Germans had played ball right up to the ceasefire. But the lateness of the German protest, as well as its directness, after a long build-up of anger with the other European allies, may have been the last straw for American patience. Addressing himself specifically to the West Germans, but hoping no doubt to be heard in other west European capitals, the US Secretary for Defence, James Schlesinger, said that if supplies in Germany could be used only for Germany and NATO and were 'not available to us for other US interests then maybe we'll have to keep them someplace else'.[63] The Secretary of State, Henry Kissinger, at one point threatened that a reappraisal of US foreign policy in Europe was on the cards. American anger was not immediately allayed when Brandt, shocked by Washington's reaction and worried about US commitment to the security of Europe, wrote a personal letter to President

Nixon, praising the Americans for their successful efforts to achieve a ceasefire and assuring them that the Federal Republic's solidarity with the North Atlantic Alliance was not in doubt.[64] For the Federal Republic it was the most serious set-back in its relations with what was its closest ally, on which it was still dependent for its security. It took a little while for the relationship to mend.

The Middle Eastern war ended after just under three weeks of fighting in another victory for Israel. As far as territory was concerned, the status quo ante was restored after the Israelis had withdrawn from the west bank of the Suez Canal. The new armistice, like the one of 1967, was the work of the superpowers which, after the cancellation of the American NATO alert, again co-operated to bring the war to an end. But this war, as has been foreshadowed, left the world in greater turmoil than its precursor in 1967, mainly because of the oil crisis and its effects, created by the continuing threat of an Arab oil boycott and the steep increase in oil prices demanded by OPEC.[65] There were fears, which turned out to be justified, that a new worldwide economic recession, already on the way before the war, would be deepened by the sharp rise in oil prices. In the shorter term the fear of the oil boycott and the fact that the Netherlands had been chosen as the only country in Europe to which it was applied, put a strain on the loyalties of West European countries and threatened the unity of the EC. The Middle Eastern countries which had participated in the fighting were licking their wounds after the most costly war since the establishment of the State of Israel in 1948. The Israeli nation had come to terms with its reverses early in the war and the resulting loss, at home and abroad, of its reputation as a virtually invincible Middle Eastern power. Above all there was now a serious rift within the Atlantic Alliance resulting from the American supply of arms to Israel.

The situation was not improved by the subsequent politics of the EC. On 6 November the nine member states of the Community issued a declaration on the Middle East intended as a 'first contribution on their part to the search for a comprehensive solution to the problem [of the Middle East]'. The gist of it was (1) that the forces of both sides in the conflict return immediately to the positions held on 22 October,[66] (2) that negotiations for a just and lasting peace be started under the aegis of the UN and according to UN Resolutions 242 and 338, (3) that peace should be based on four principles: the inadmissibility of acquiring territory by force, the ending of the occupation of the territories held by Israel since 1967, respect for the sovereignty and territorial inviolability of each state and its right to live in peace within secure and recognised borders, and a recognition of the legitimate rights of the Palestinians, and (4) that a settlement must be reinforced by international guarantees which must be strengthened by an international peace-keeping force established within the demilitarised border zones envisaged by Resolution 242.[67] This declaration therefore repeated the points alleged to have been made in the unpublished 1971 text to which Israel had strongly objected, and even went further. The first point could only be directed against the Israelis who were then still on the west bank of the Suez Canal surrounding Suez and the Egyptian Third Army. But the point to which Israel took the greatest exception was the third one: the implication that Israel had acquired territory by force and had not been mentioned by name as having a right to live in peace within secure and recognised borders, since it was Israel which had been denied this right for 25 years by its Arab neighbours. As for the last point, the declaration again raises the contentious demand, refused by Israel and which was not part of Resolution 242, of a peace-keeping force along the borders. This new EC declaration was roundly condemned by Israel as being unashamedly pro-Arab. Many voices in the West, including the Federal Republic, supported this view. The reason why it was made was clearly to protect Arab oil supplies.

What came as a great surprise was that Chancellor Willy Brandt had signed the declaration on behalf of his government. While there are lingering doubts about who signed what in 1971, there was no doubt this time. It seemed a long way from the warm reception and friendly talks only five months previously in Israel and Brandt's statements that his government's interest was 'the achievement of a peaceful solution negotiated by and acceptable to the parties directly involved', an idea noticeably absent from the EC declaration. Neither did the declaration appear to accord with Brandt's speech to the press in Jerusalem, when he said that he would bear Israel's economic and political interests in mind when the newly extended Community set about its task of determining Mediterranean policy. But what was even more surprising was Brandt's reply when he was challenged about the declaration in Israel and by Israel's many supporters in the Federal Republic. Addressing the Bundesrat on 9 November, he maintained that there was no change in West German attitudes and policies and that the Federal Republic's relations with Israel were also unchanged. He again referred to them as being of 'a special character'. But he added that the 'accent set' at EC level may have been slightly different as it was necessary to consider one's partners. The EC resolution had been a compromise which gave some dissatisfaction to everyone. It was the price for the desired unity of Europe. It was ultimately in the interest not only of the Europeans but also of the partners of the Middle Eastern conflict that the EC should gain in importance. Brandt also said that the declaration had not finalised anything. It should not be regarded as one-sided, and the Federal Republic stood by the policy of the EC such as it had been developed.[68] The gist of this statement was repeated to the Bundestag.

It could be that the Federal Government felt itself under such strong pressure because of the threat of an Arab oil boycott that it needed to support the EC declaration. However, that is hardly borne out by the general atmosphere in the country. It is true that petrol rationing was briefly introduced and restrictions on driving were enforced for a time in the Federal Republic, as happened in most other west European countries. This caused some unease among the public but judging by the government's statements and press reports there was no panic since the Federal Republic was not immediately threatened by the boycott and stocks were said to be adequate for some time. It was stressed that OPEC was not united over the boycott and that one of the largest producers, Iran, had refused to join it. Even if the boycott was extended, few people believed that the Arabs would be in a position to keep it up for long.[69] Be that as it may: if fear for Arab oil was a pressing reason for supporting the EC declaration, why virtually disown it three days later? Once again it seems that the text of the declaration was prompted by the French government and supported by other member states of the Community with similar policies. There is no obvious link here with the state of Franco-German relations as there was in 1971, since the difficulties of that time had been overcome. But some show of European solidarity was undoubtedly thought appropriate in view of the general uncertainty and feeling of insecurity in Europe, not only with regard to the Arabs but also because of the deteriorating relations with the US, still the indispensable protector of western Europe. But it is difficult to see how a declaration such as the one passed by the EC, arrived at in this way, could make a strong impression, let alone a lasting impact. Once again the conclusion must be that such an act of 'solidarity', if not based on real consensus, only highlights the differences between the individual member states. Like the 1971 report it had no sequel. It did not set the scene for a new united and consistent EC policy in the Middle East, nor did the Federal Republic at this stage determine its policy in accordance with the will of the Community. Such influence as could be exerted on the Middle East from outside was exerted by the superpowers.

Criticism in the Federal Republic of the way the government handled the crisis was muted. This may be a reflection of the concern with which the West German public was facing the energy crisis. What criticism there was came mainly from the right of the political spectrum. The opposition leader in the Bundestag, Karl Carstens, CDU, accused the EC foreign ministers of an 'insufficient sense of proportion'. The declaration, he said, did not take Israeli interests sufficiently into account.[70] A number of

CDU/CSU deputies protested to the leadership of their parliamentary parties that the declaration was not neutral but clearly supported the Arab side. They charged that the desire to meet the threatened shortages of Arab oil deliveries and the resulting difficulties evidently weighed more heavily than moral considerations and the responsibility for the life interest of a state.[71] The Springer press, traditionally very pro-Israel, was more outspoken. A leading article in *Die Welt*, in a reference to the EC governments, read:

> In a major world political crisis they have abandoned every appearance of neutrality, supported the Arab point of view and taken sides against Israel and in practice against the USA. ... Great Britain, France and Italy may feel morally free to make such decisions. But when the Federal Republic, of all states, thereby enables the first joint European decision on international questions to take sides against Israel in an Israeli life-or-death matter, then that must make every German pale with disgust.[72]

The *Frankfurter Allgemeine Zeitung* referred to the Arab oil weapon as an instrument of blackmail and thought that at a future Middle Eastern peace conference the EC declaration would cause the Europeans difficulties.[73] But the EC declaration was overshadowed by the crisis in West German relations with the USA. The chairman of the CDU, Helmut Kohl, referring to the ban on US arms to Israel from German territory, said that the Federal Government had 'placed itself outside the solidarity of the Western Alliance' by the ban.[74] The *Frankfurter Allgemeine Zeitung* wrote that Bonn had put months of exemplary co-operation with the US at risk and asked what was more important: 'our dependence on the Arabs for oil or on the US for defence?'[75]

On the Israeli side there had at first been no official reaction to the West German arms ban. A foreign office spokesman refused to comment, saying that this was an issue between the US and Germany.[76] In connection with this the *Jerusalem Post* asked:

> Did Israeli restraint result from a recognition that Bonn's objections were made only when the fighting was over and it became public knowledge? Israeli officials do not discount the possibility that the Bonn government in fact turned a blind eye to the arms supply until it was embarrassed into action and even then acted only after the cease-fire had already been announced.[77]

But in a radio interview, the Israeli ambassador in Bonn, Eliashiv Ben Horin, accused the Federal Republic of giving in to Arab blackmail. Brandt, he said, had often stressed Germany's special relationship with

Israel and its right to live. He questioned how this could be consistent with a policy of strict neutrality at a moment when Israel was fighting for its life. The existence and security of Israel could not be made to depend on whether the central heating in other countries functioned. Those who wanted only for themselves to sit by a warm fire should be ready for surprises: blackmailers were never satisfied. It had not been a football match but his country's gravest fight for its existence.[78] The Israeli evening paper, *Ma'ariv*, commented on the West German arms ban: 'This is neutrality which indirectly contributes to genocide'.[79] When the EC declaration was published there was more condemnation. The Israeli Ambassador to the EC, Moshe Allon, said in Brussels: 'This declaration espouses the Franco–Soviet thesis, i.e. the Arab position, in an illusory hope of guaranteeing oil supplies'.[80] The Israeli Prime Minister, Golda Meir, attacked the Europeans for their 'Munich attitude' towards Israel, saying that Chamberlain's policy had not helped in the long run.[81] On 11 November Brandt, concerned about the fierce Israeli reaction, said in an interview on West German radio that the EC declaration had not blocked the way to some territorial adjustments; and the day before he had hurriedly sent his second-in-command in the SPD, deputy chairman Heinz Kühn, to Jerusalem to smooth matters over and to express 'his solidarity and sympathy'.[82]

Brandt's action showed that Israeli comments on the policy of his government still mattered. Even though, as was shown during the October war, popular attitudes to Israel were no longer quite what they had been in the Federal Republic in the past, Israel was still a factor in the search by the West German nation for redemption from the burden of the Nazi era. That Israel had to survive in the latest Middle Eastern war was of paramount concern to the government and people. There was from Israel's point of view a bright side: that at a crucial moment for Israel the Federal Government, almost alone among the large West European states, continued to connive at the airlifting of American arms from NATO stores in West Germany right up to the end of the fighting and indeed beyond the armistice. On the other hand, the period described in this chapter produced evidence of changes in the policy of the Federal Government towards Israel. The question of Israel's survival apart, Israeli interests were taken into account less when the West German national interest was at stake. The change of the phrase 'special relationship' to 'normal relations of a special character' used by the Federal Government in describing its relationship with Israel was not merely cosmetic. It emphasised the 'normal' aspect, the intention of henceforth treating Israel like other states, with the addition, however, that the past must not be forgotten, even less condoned. In practice this meant that the Germans would no longer go out of their way just to suit the Israelis. There remained, for example, in West Germany's Middle Eastern policy the dilemma between its relations with Israel and with the Arabs. There were indications that the Federal Government was no longer willing to put at risk the newly regained relations with the Arab states for the sake of Israel where interests clashed. It was not only Arab oil that was involved. What is also important is that the West Germans had regained a measure of international importance and freedom of action which made it possible for them, more than at any other time since the Second World War, to pursue their own independent policies in accordance with what they perceived to be their interests.

NOTES

1. Meaning two German states. Since 1955 the USSR had insisted on Germany remaining divided.
2. Interview with the Lebanese newspaper *Al-Hayyat* on 28 June 1969: see *Deutschland Berichte*, July 1969.
3. See Christopher Imhoff: 'Die Konstruierte Krise' in *Tribüne*, vol. 9, no. 33, 1970.
4. Suspicion seems to have arisen because of pro-Israel statements by Brandt, followed by pro-Arab statements by Scheel, designed to preserve the 'balanced' West German approach to the Middle Eastern conflict. That these constituted differences of opinion between two politicians was emphatically denied in Bonn. *Deutschland Berichte*, March 1973.
5. The best example was Gerhard Schröder, Federal Foreign Minister, 1961–66.
6. Statements that Israel was 'a state like any other' and that relations must develop accordingly were made by the Federal Foreign Office in summer 1971. Scheel, when challenged, countered these by emphasising the 'special German–Israeli relationship' (see *Die Welt*, 9 July 1971).
7. *Deutschland Berichte*, July 1969.
8. Deligdisch, p. 147.
9. On 28 June 1969. Quoted in *Deutschland Berichte*, July 1969.
10. The French translation refers to 'withdrawal from "the" [i.e. all] occupied territories'; the British insisted on 'withdrawal from occupied territories' leaving open the possibility of border changes. Arab and Soviet Bloc states recognised the French, the UK, Federal Republic and Israel recognised the English translation.
11. *Die Welt*, 15 & 16 May 1971.
12. A copy of a first draft, according to *Der Spiegel* (12 July 1971, p. 25) was 'obtained' by the Israeli secret service; according to the *Jerusalem Post* (12 May 1971) it was 'leaked to the press'.
13. Peter Cycon in *Die Welt*, 10 & 11 July 1971.
14. The other three countries were Denmark, the Irish Republic and Norway. Norway withdrew after an internal referendum.
15. The only official statement issued by the EC foreign ministers during their deliberations reiterated support for UN Resolution 242 saying 'we affirm our approval of the Resolution which constitutes the basis for a settlement'. In view of the conflicting interpretations this statement did not clarify anything.
16. *Die Welt*, 21 May 1971.
17. Ibid.
18. *Jerusalem Post*, 11 May 1971.

19. *Jerusalem Post* 18 May 1971.
20. Ibid, 12 May 1971.
21. Ibid, 19 May 1071.
22. *Der Spiegel*, 21 July 1971.
23. *Die Welt*, 10 & 11 July 1971.
24. Two were shot during the first attack, nine were taken hostage and killed when German police tried to free them by force.
25. *Der Spiegel*, 6 Nov. 1972.
26. Ibid.
27. Ibid (quoted).
28. *Der Spiegel*, 6 Nov. 1972.
29. This may have been reflected in an opinion poll taken 7 months later by the Allensbach Institute, which recorded that only 37% of the population were pro-Israel.
30. The German adjective 'ausgewogen', used frequently since the Six Day War, is difficult to translate precisely but conveys the idea of 'balance'.
31. Brandt had some time before knelt before the memorial of the Warsaw Ghetto when visiting Poland. It was an unprecedented gesture that received worldwide publicity and made a deep impression in Israel.
32. Cf 'In God's eyes the man stands high who makes peace between men . . . . But he stands highest who establishes peace among nations.' (*Talmud*.)
33. Golda Meir at Knesset reception for Brandt, 7 June 1973, *Europa Archiv*, no. 14, 1973, p. D389 (Dokumente).
34. Willy Brandt at Knesset reception, 7 June 1973, *Europa Archiv*, ibid.
35. Press Conference in Jerusalem, 8 June 1973. See *Europa Archiv*, no. 14, 1973, p. D390 (Dokumente). The words 'normal relations of a special character' should be noted. Their meaning in the eyes of the West German government is different from the original 'special relationship' still adhered to by the Israelis.
36. On 9 June 1973. See *Europa Archiv*, no. 14, 1973, p. D391 (Dokumente).
37. *Jerusalem Post*, 21 June 1973.
38. *Ha'aretz*, quoted in *Jerusalem Post*, 21 June 1973.
39. *FAZ* 12 June 1973.
40. *Ma'ariv*, quoted by *FAZ*, 12 June 1973.
41. Jordan did not participate this time. Iraq and Saudi Arabia sent small contingents.
42. The '*Grundlagenvertrag*': the most important achievement of *Ostpolitik*, signed in Feb. 1973.
43. Deligdisch, p. 188
44. Ibid.
45. The Netherlands' policy towards Israel was little different from that of the Federal Republic, but because Rotterdam is a large oil distribution centre for Northern Europe it may have been chosen to give the boycott the greatest possible effect.
46. *Der Spiegel*, 29 Oct. 1973, pp.25–28.
47. *Daily Telegraph*, 26 Oct. 1973.
48. Ibid.
49. *Daily Telegraph*, 27 Oct. 1973.
50. *Daily Telegraph*, 26 Oct. 1973.
51. After agreement had been reached between Brezhnev and Kissinger in Moscow a UN Security Council resolution was passed on 22 Oct. that military operations should cease in the evening of the 23rd. But Israeli units crossed the Suez Canal on the 24th and surrounded Suez and the Egyptian Third Army. Apart from sporadic shooting in that sector fighting had ceased on the 25th.
52. *Jerusalem Post*, 5 Nov. 1973.
53. *Der Spiegel*, 15 Oct. 1973, pp. 25–27.
54. *Le Monde*, 13 Oct. 1973.
55. *The Times*, 20 Oct. 1973.
56. *Deutschland Berichte*, Sept./Oct. & Nov. 1973.
57. *Die Welt* 24 Oct. 1973.

58. *Frankfurter Rundschau*, 26 Oct. 1973.
59. *Jerusalem Post*, 26 Oct. 1973.
60. *Daily Telegraph*, 26 Oct. 1973.
61. 2.2 million American servicemen around the world were placed on 'Defcon 3' which means enhanced readiness. The reason was fear the USSR might send troops to reinforce the Arab armies. It was revoked the following day (see *The Times*, 27 Oct. 1973).
62. *Daily Telegraph*, 26 Oct. 1973.
63. *International Herald Tribune*, 29 Oct. 1973.
64. *Südd. Ztg.*, 31 Oct. 1973.
65. The price of oil was increased by OPEC by 70% on 16 Oct. 1973. By 6 Nov. 1973 Arab oil production had been cut by 25%: further monthly cuts of 5% were envisaged (see *Jerusalem Post*, 5 Nov. 1973 and *Südd. Ztg.*, 6 Nov. 1973).
66. The date of the Security Council Resolution requiring a ceasefire.
67. See *Europa Archiv*, sequence 2, 1974, p. D29 (Dokumente).
68. *Deutschland Berichte*, Dec. 1973.
69. See *FAZ*, 18 Oct. 1973.
70. *Jerusalem Post*, 7 Nov. 1973.
71. Press archives of *Deutsch-Israelische Gesellschaft*, Bonn.
72. *Die Welt*, 7 Nov. 1973.
73. *FAZ*, 15 Nov. 1973.
74. Deligdisch, op. cit., p. 190.
75. *FAZ*, 29 Oct. 1973.
76. *Jerusalem Post*, 26 Oct. 1973.
77. *Jerusalem Post*, 5 Nov. 1973
78. *Die Welt*, 6 Nov. 1973 and *Jerusalem Post*, 7 Nov. 1973.
79. *Ma'ariv*, quoted in *Frankfurter Rundschau*, 27 Oct. 1973.
80. *Jerusalem Post*, 7 Nov. 1973.
81. *Jerusalem Post*, 12 Nov. 1973.
82. *Deligdisch*, p. 191.

11

The Troubled 1980s

Israeli anger aroused by the Federal Government's arms ban and the European Community declaration soon calmed down. Realisation in Israel that the Federal Republic, despite appearances, had given important aid by allowing the American airlift from its territory undoubtedly contributed to the calmer atmosphere. In effect, relations returned to normal, that is according to the latest West German interpretation, but not to where they had once been and where Israel would have liked them to have remained. But that point had been passed several years ago. Three decades had now elapsed since Auschwitz. Time was bound to mitigate any feelings of guilt for what had happened then. But there have been no serious incidents to disturb the West German–Israeli relationship since the end of 1973, yet mainly because of Israeli sensitivity there has been a certain amount of friction between the two countries. Changes of personalities exacerbated this phenomenon for a time. In 1974 Willy Brandt resigned from the chancellorship and was replaced by Helmut Schmidt, again of the SPD. Despite recent Israeli criticism there was much good will for Brandt in Israel as a man of great moral stature, prompting one newspaper to write of him that to Israel 'he has always shown sincere friendship ... We want to remember Brandt as the one who knelt before the Ghetto memorial ...'[1] In 1977 the cabinet of Golda Meir was replaced by the right-wing Likud government of Menachem Begin, the leader of the Herut party which had persistently opposed any dealings with the Germans. Begin now accepted the German connection while Brandt's successor Schmidt also continued broadly his predecessor's Israel policy. But the more abrasive nature and authoritarian style of the one and the occasional return to the highly emotional and hawkish statements of the other, remembered from his time as Leader of the Opposition in the Knesset, led to angry exchanges and ruffled feelings on both sides.

Basically the assessment made in 1971 about the West German attitude to Israel was still true and remained so throughout the 1980s: that German responsibility for the murders of the past was not denied but must no

192

longer restrict the freedom of the West German government in the
political field. In various forms this was indeed spelled out by members of
the Federal Government on several occasions: by Chancellor Schmidt
when he told Begin that, while he agreed with him that the Germans
should have a conscience about the past, this was no reason to support
Israel;[2] and by Chancellor Kohl, Schmidt's successor, when he said he was
conscious of German responsibility for the fate of the Jews but would
on no account support all decisions of the Jerusalem government.[3] In
practice this meant that, for example, the West Germans would give
support to Palestinian self-determination and would express opposition
to the building of Jewish settlements in the Israeli occupied areas – which
angered the Israelis – yet would always insist that Israel had a right to live
in peace within secure and recognised borders. The West Germans were
still reluctant to support any measures that would imperil Israeli security.
About what this meant exactly there might be controversy as the view was
gaining ground in the West that Israel, the accepted military champion in
the Middle East, was overstating its security needs. Generally, however,
the period of material assistance by the Federal Republic to Israel being
now more or less over,[4] the relationship became characterised more by
statements and speeches on both sides. It was therefore in words rather
than in actions that, in the eyes of the Israelis, the West Germans failed
the Jewish State.

In a sense the Federal Government was still searching for evenhandedness in its relations with both Israel and the Arab states. But this
became distorted by the developing oil crisis and new upheavals in the
Middle East. In 1979 the Shah of Iran was ousted by revolution and the
country became an Islamic republic under the Ayatollah Khomeini. In
the same year the Soviet Union invaded Afghanistan and this was followed
in 1980 by the Iraqi attack on Iran leading to the first 'Gulf War'. Since the
mid-1970s a civil war had been raging in Lebanon between the Maronite
Christians who held most of the power and the Muslims who wanted a
bigger share of it. Israel, concerned about increased military activity of
Palestinian refugees in the war-torn country, decided to invade. Over-
hanging all was the oil crisis, deepening because of further cuts in
production by some Arab states and more price increases. The only
positive events were the peace treaty between Israel and Egypt, leading
to the Israeli evacuation of the Sinai peninsula, and the Camp David
Agreement which was to have affected the situation of the Palestinians in
the Israeli-occupied areas. Both were merely the first steps towards a
general peace settlement, negotiations for which were started only in
1993 and at the time of writing are still continuing.

The invasion of Afghanistan raised new fears of Soviet expansionism, especially among the small and medium-sized states adjoining the Arabian Gulf. It has already been said in this book in another context that the Soviet action again deepened the East–West conflict and led to a more militant anti-Soviet policy in Washington. What is of interest here is that the new threat led to efforts by the Gulf States, one of the largest and most important of which is Saudi Arabia, to rearm in order to meet this danger. The weapons obviously had to come from the West, including the industrially developed countries of western Europe. But these countries were now fearing for their oil supplies, most directly because of the artificially created oil shortage, more remotely by what seemed a threat to the whole of Middle Eastern oil because of the Soviet invasion of Afghanistan and the war between Iraq and Iran, both major oil producing states in the region. The western European states in question, most of which were members of the European Community, now had to be particularly careful how they handled their relations with the Arab states and Israel. Most of them had been pro-Arab before, mainly because of oil interests. They needed Arab money because of increasingly negative balances of payments with the Arab states resulting from high oil prices. One way of redressing this was through Arab investments in Europe, the other by increasing European exports to oil rich countries in the Middle East. What these countries wanted to buy most was military hardware. But large sales of sophisticated weapons to Arab countries in turn created an increased feeling of insecurity in Israel.

This, broadly speaking, still was the situation during the early and middle 1980s. It brought to the fore once again the dilemma in Bonn of the Federal Republic's relations with Israel and the Arab states. To the political and economic factors which increased this dilemma should be added some internal ones. German sympathy for Israel was further declining. There are numerous reasons for this: there was increasing sympathy for the Palestinians living on the occupied West Bank of the Jordan and in the Gaza Strip. An occupation by a foreign nation is never 'benevolent' for those who live under it and do not desire it. Politicians used the term 'Palestinian self-determination', although it was never spelled out precisely what this meant. Did it require an independent Palestinian state in the areas of present Israeli occupation, or could, if the situation demanded it, an autonomous region under Israeli sovereignty provide an acceptable alternative? West German politicians linked it to German self-determination, the right of all Germans, including those then living under communist rule in the German Democratic Republic, to live as free, democratic citizens, preferably in a reunited Germany. It is a poor comparison, given the totally different circumstances in the Middle

East but served as a useful gimmick for Federal Government members to explain to the angered Israelis and sceptical Arabs their reasons for their attitude. Linked to the rights of the Palestinians was the Israeli policy of creating Jewish settlements in the occupied areas. These were justified by religious hard-liners by the right of Jews to live anywhere in the ancient biblical land of Israel, by the government mainly as a security measure; but many people outside Israel saw it as an attempt to 'create new facts' by settling large numbers of Jews in the area in order to justify its ultimate incorporation in a 'Greater Israel'. The building of these settlements was therefore opposed by the Palestinian population, fiercely attacked by Arab governments and generally criticised in most other countries.

After the European Community declaration of 6 November 1973 which first mentioned the 'legitimate rights of the Palestinians' other such declarations and statements by various governments multiplied, becoming stronger and more explicit on the question of Palestinians as time went on. They soon included the phrase 'self-determination of the Palestinians' and later explained that this could mean some political entity, possibly but not necessarily an independent Palestinian state on the West Bank of the Jordan and in the Gaza Strip. All European Community declarations were, of course, made in the name of all the member states, including the Federal Republic. This development must be ascribed to the energy crisis of the late 1970s and early 1980s, that is, the fear of the western European states for their oil supplies from Arab sources. It has been suggested, and this is very likely, that the Federal Government helped to tone down the wording of the more militantly pro-Arab texts at first proposed by some member states, especially France. Thus the European Community declarations became sufficiently vague to allow for different interpretations and included a mention of Israel by name as one of the countries entitled to live in peace within secure and recognised borders.

It would be futile to go into detail about the many statements of the European Community but points from two declarations should be mentioned. One was made by the European Council on 29 June 1977. After acknowledging that all states in the region have a right to live in peace within secure and recognised borders and that the Arab side must accept this in the case of Israel it goes on to say that a Middle Eastern settlement should take account of the need for a Palestinian homeland and that representatives of all the parties to the conflict should participate in the negotiations.[5] There is no mention as yet of an independent state nor of any role of the Palestine Liberation Organisation. More important and more comprehensive is the declaration issued by the European Council in Venice on 13 June 1980. In this the nine member states say that the time has come to:

implement the two principles universally accepted by the international community: the right to existence and security of all states in the region, including Israel, and justice for all the peoples, which implies the recognition of the legitimate rights of the Palestinian people.

They go on to state that the Palestinans must be placed 'in a position where they can exercise fully their right to self-determination'. They do not elaborate. They also demand that the Palestine Liberation Organisation be associated with the negotiations, say that they will not accept any unilateral initiative designed to change the status of Jerusalem and that the Israeli settlements constitute a serious obstacle to the peace process. Finally they envisage 'a system of concrete and binding international guarantees' for a peace settlement to be provided by the United Nations and in which the nine member states offer to participate.[6]

If this was regarded by the European Community heads as the start of an initiative it certainly failed in its purpose. All that did happen was that it annoyed the Americans and, of course, the Israelis. The United States government, in its efforts to advance the peace process, had taken roughly the Israeli line as expressed in the Camp David Agreement and which did not coincide with the line taken by the European Community at Venice. Washington may well have considered the European Community action inopportune and harmful to the attempts it was making towards a solution of the Middle Eastern conflict. The *New York Times* remarked scathingly:

> The notion of Europe playing a role as an effective third force remains just that, a persistent notion but not a reality. The Common Market nations lack structural cohesion to play an effective role ... As a declaration of independence from American diplomacy in the Middle East, the European allies' pronouncement in Venice ... was merely pathetic ... a petulant cry that Europe wants to play big-league power too ... a cramped call for a 'comprehensive' solution, their code language for a denunciation of Camp David.[7]

But the Venice Declaration, like previous attempts by the European Community to intervene in the Middle East, had no effect and only demonstrated once more that if the superpowers could do little in the region, the Europeans could do nothing. There was still barely a consensus in the European Community and little attempt to see both sides of the problem.

Israel was predictably very critical of the Venice Declaration though some Israeli politicians claimed to have been privy to an earlier draft which from their point of view was even more unfavourable and was then changed by pressure from the Germans. An unofficial reaction in Jerusalem was: 'Bad, but it could have been worse'. It did not recognise the Palestine Liberation Organisation as representing all Palestinians nor tamper with UN Resolution 242 except for a reference to refugees. The offer by the European Community of security guarantees was referred to as 'interesting but unacceptable'.[8] Apart from that most points were rejected. The Israelis had never accepted the idea of Palestinian 'self-determination' as they suspected that this meant a Palestinian state which they totally rejected, as they rejected participation of the Palestine Liberation Organisation. Since the government and Knesset had recently passed a law officially annexing the formerly Arab East Jerusalem to the Jewish State, against which the reference to Jerusalem in the Declaration was clearly a form of protest, this reference too was a source of annoyance. Generally the Venice Declaration was regarded as an attack on the Camp David Agreement, which had been totally rejected by the Arab states except Egypt, and by the Palestine Liberation Organisation. The Prime Minister, Menachem Begin, who was wont to react in times of crisis with comments which were emotional and often irrelevant, in a cabinet statement called the declaration 'a Munich-like surrender to tyrannical extortion' and as 'seeking to destroy the peace process'. His more rational Foreign Minister, Yitzhak Shamir, commented that 'nothing will remain of the Venice Declaration but its bitter memory'. He claimed that Israel had made major concessions while the other side had made none.[9] The *Jerusalem Post*, in a leading article, wrote:

> Fearful of what Arab oil states might do to their creaking economies, yet hopeful of establishing themselves as a world force vis-à-vis the Americans, the Economic Community damned Egypt–Israel peace and the Palestinian autonomy negotiations with studied inattention.[10]

whether they were addressed to an Arab or an Israeli audience. In practice this mattered little because they were sufficiently vague and there was not much chance that the advice given in them would be heeded, since such initiatives as were taken were handled by Washington. The points made and often repeated in Bonn or elsewhere by Federal Government ministers were by and large that: the Palestinians have a claim to realise the right to self-determination; they have a right to a homeland; they have the sole right to choose whether they should have an independent authority in the territory vacated by Israel or would accept another solution; they have the sole right to decide who should represent them in the negotiations. These were not approved of by Israel because they went further than the Camp David Agreement which envisaged only a form of autonomy in the territories in question under Israeli sovereignty. The reason why the West German and European Community demands were unrealistic was that they did not take into account the disturbed situation in the Middle East and in particular the fact that of all Israel's Arab neighbours only one, Egypt, had made peace with Israel and issued a declaration of intent that it accepts and recognises the Jewish state and will not go to war with it again. To this must be added that the PLO, which claims to represent all the Palestinians in the occupied areas, was dedicated, like some of Israel's neighbours, to Israel's destruction and was not, and is not today, an independent organisation with a unity of purpose. When one considers what would be an independent Palestinian state which might well reflect the disunity of the Arab world as that disunity is reflected within the PLO, it is indeed difficult to imagine how Israel's security could be assured. From occasional unofficial asides made by West German politicians one receives the impression that they understand this problem but that the government cannot deviate from the adopted line for political reasons. One may also conclude that if relations between Jerusalem and Bonn have been and still are friendly it must be because the Israeli government later, the Israeli invasion of Lebanon. But the disputes that arose were sparked off mainly by Begin's emotional and flamboyant style and Schmidt's impatience and brusqueness. What had caused irritation in Israel was the apparent flirtation of the Federal Government with Saudi Arabia, which was technically still at war with Israel, during the late 1970s and 1980s. There were exchanges of visits by members of both governments. Oil and financial arrangements were the inducements for the Germans and as usual the benefits exacted a political price. A statement issued jointly at the end of a visit by Crown Prince Fahd to Bonn said that the Palestinians had a legitimate right to their own state and Schmidt described the Israeli response to the recent peace visit by President Sadat of Egypt to Jerusalem as 'inadequate'.[13] Ma'ariv commented bitterly: 'Germany does not want to lag behind other states of the world in its quest for the money and oil of Saudi Arabia, even if the cost is demanded in Israeli currency'.[14] Schmidt had gone beyond the declaration made up to that time by the European Community and his own government by spelling out the need for an independent state. Moshe Dayan, the Israeli Foreign Minister, commented during a visit by the Federal Foreign Minister Hans-Dietrich Genscher to Jerusalem which followed the Saudi Prince's visit to Bonn:

> 'We expect you, after all that was done to the Jewish people in your country and in Europe during the Hitler period, always to pay special attention to the security of this people and this country. A Palestinian state cannot be combined with the security of Israel as it will one day serve as a spring-board for an attack against Israel.[15]

There were several such statements made by Chancellor Schmidt during his period of office ranging from self-determination for the Palestinains and the role of the PLO in the peace process to criticism of the Begin government's settlement policy in the occupied territories. There was also personal innuendo on both sides. In the autumn of 1980 Schmidt said of Begin that he was 'a danger to peace'. In return Begin said about Schmidt: 'Maybe he, as an officer at the front had himself participated in atrocities against Jews'. This accusation was false: Helmut Schmidt is considered to have a completely clear wartime record. It is believed that after this attack Schmidt, who had accepted an invitation to visit Israel, decided not to go as long as Begin was Prime Minister.[16] But the most serious verbal clashes occurred after Schmidt's visit to Saudi Arabia in the spring of 1981. In a television interview following his return to West Germany the Chancellor said that Bonn had 'a moral commitment to the Palestinians who fled from the West Bank'. There was apparently no mention of the Germans' commitment to the Jews and to Israel. He further made it clear that he was reviewing the Federal Government's policy not to sell arms to areas of tension and called for inclusion of the PLO in the peace negotiations. He said of the Palestinians' right of self-determination that 'for me this includes their right to organise themselves as a state', before mentioning that Israel's borders must be recognised. Finally he referred to the Saudis as the Federal Republic's chief allies – both politically and economically – outside Europe and the United States. Prime Minister Begin, addressing the executive of the Herut Party of which he was the leader, called Schmidt a hypocrite and accused him of wilfully forgetting the German crimes against the Jewish people and of being interested only in selling weapons and buying oil cheaply. He added: 'There is no remnant of heart, humanity, morality or memory ... It seems that the Holocaust had conveniently slipped his mind ... We hear of a commitment to those who strove to complete what the Germans had started in Europe'. Later he said: 'Schmidt does not care if Israel goes under. He saw this almost happen to our people in Europe not so long ago. He served in the armies that encircled the cities until the work was finished by the Einsatzgruppen'.[17]

This kind of personal insult certainly angered not only Schmidt but also the German public. Nevertheless the West German authorities on the whole reacted calmly. Begin's speech, even if some resentment was understandable, had clearly been unnecessarily strong and had therefore had little effect. The West German government spokesman, Kurt Becker, thought Begin's reactions might be linked to the Israeli election campaign, then in progress. When asked about German–Israeli relations he said they were normal but only if no account was taken of Begin's outbursts.[18] The West German press was almost universally hostile to Begin's speech and supportive if perhaps not uncritical of the Chancellor.

The quarrel blew over after a time but the question of possible West German arms deliveries to Arab countries, back on the agenda when Schmidt made his statement about changes in German export regulations, remained a bone of contention. There had been talk throughout most of 1978 about rockets of West German manufacture being exported to Syria. But as Genscher, on a visit to Israel, pointed out these weapons were in fact co-produced by the Federal Republic and France. While Bonn did not then sell arms to areas of conflict the French had no such policy and the Federal Government could not veto French exports of jointly produced goods.[19] But the Leopard II tanks, manufactured solely in West Germany and regarded as the best in the NATO alliance were another matter. Saudi Arabia became very interested in these following the Soviet invasion of Afghanistan and western European states were anxious at that time to export military equipment to the Middle Eastern states for the reasons already indicated. It was claimed in the West, including the USA which participated in these exports, that they would not serve any other purpose than the defence of the Gulf region against any encroachment by the USSR or possibly the new Islamic Republic of Iran. This view was not shared by the Israeli government, a spokesman of which asserted that such weapons would then be passed on to the enemies of Israel or used against it. He emphasised that in all the wars against Israel so far Saudi Arabia had directly participated with troops, money and weapons. It was wrong therefore for the Federal Republic to take part in the arming of the Arabs.[20] Nevertheless the Bonn Government, in May 1982, had abolished the restriction on arms exports to areas of tension, as Chancellor Schmidt had envisaged. Though, as *Der Spiegel* at the time commented, the government would continue to exercise restraint, the door was in theory wide open to arms exports, including 300 Leopard II tanks requested by Saudi Arabia and submarines to Chile. The reason given by the magazine was 'a recession in the armaments industry'.[21]

By January 1984 when Chancellor Helmut Kohl, who had replaced Schmidt two years earlier, visited Israel the matter had not been resolved. A few days before his departure, when challenged by the Israelis, a Federal Government spokesman assured Jerusalem that the problem of the Leopard II tank was 'off the table'.[22] After Kohl's return it seemed to be on again when he told the Saudis that export of West German arms to them was conditional on these not being used against Israel. Three months before, when in Saudi Arabia, Kohl had apparently offered them without such conditions. But the matter was again put on ice. It was still there in October 1985 when the President of the Federal Republic, Richard von Weizsäcker paid the first ever visit by a West German head of state to Israel. But by this time the Leopard tanks had been joined by a high-technology munitions factory which the Saudis had ordered to be built in their country by the West German firm Rheinmetall but which was not subject to the same export controls as actual weapons. This was stated during a Bundestag debate on 17 October in which there was severe criticism of West German arms exports to the Middle East by members of all parties. The head of the Chancellor's Office, Wolfgang Schäuble, replied that an understanding with Saudi Arabia was 'an expression of our vital interest in the stability of the Gulf region' and that this interest was shared with all other western states, especially the USA, France and Great Britain. But his government was still considering the matter.[23]

The problem was still unresolved in April 1987 when a new controversy erupted over arms exports generally and to Saudi Arabia in particular. It was played down because of the impending return visit by Israeli

President Chaim Herzog. Ironically it was Franz-Josef Strauss who was one of the protagonists of these exports. In the *Bayernkurier* he had claimed that it was in Israel's interest that 'moderate Saudi Arabia' should be made strong as this would have a 'stabilising effect'.[24] But by November of that year the export of heavy German weapons like the Leopard II tanks had been quietly dropped, according to the *Frankfurter Allgemeine Zeitung*, 'above all because of pressure by pro-Zionists in the Federal Republic'. It had persistently been claimed that the tank and other war equipment could be used against Israel.[25] Who the 'pro-Zionists' were was not explained but Israeli protests had clearly found an echo among West German members of parliament and the public. The government seems to have given way to public pressure.

The foregoing may give the impression that relations between West Germany and Israel during the 1980s were bad. But the mutual recrimination between Begin and Schmidt and the disagreement over arms sales to Arab states, though they momentarily worsened the atmosphere, were only one side of the relationship. The Israeli government realised from the outset that the Federal Republic was far more reticent over arms deliveries to Arab countries than any other western industrialised state, even if it expected special consideration from Bonn. On the other hand Israel did receive assistance from the Federal Government in obtaining a fair deal from the European Community at a time when two countries, Greece and Spain, both potential competitors of Israel in citrus fruit and other Mediterranean produce, were about to join the Community. Moreover, trade with and West German investments in Israel flourished and there was much in the way of voluntary German contributions, public and private, to scientific, social and cultural institutions in Israel as well as contacts among groups and individuals. The personal animosities between two heads of government do not have to determine the underlying relationship of the two countries. In this case they simply highlight the delicacy of this relationship which is so deeply rooted in the events of the past.

A truer picture of Israeli–German relations is perhaps given by the two heads of state who came on official visits to each others' capitals in the mid-1980s when Schmidt had been replaced as Chancellor by Helmut Kohl and Begin as Prime Minister by Yitzhak Shamir. Both presidents, Richard von Weizsäcker in Germany and Chaim Herzog in Israel are constitutionally limited in the exercise of political power in their respective countries.

This could reduce their visits to a mere formality. On the other hand, because of the prestige of their office and precisely because they are deemed to be above politics and are both men of high intellectual and moral calibre, they may reflect the underlying relationship better than high ranking politicians. Weizsäcker went to Jersualem in October 1985 and Herzog returned the visit in April 1987. It was the first time such a visit had taken place by a West German or Israeli head of state. In their speeches both emphasised that German youth must not be made responsible for the past but for what emerges from it in the future. It was to the future therefore that both countries should address themselves.[26] Herzog said when welcoming Weizsäcker in Jerusalem: 'The past which determines both our present and our future hovers between our two peoples like an invisible wall and yet we know only too well that the future cannot be determined exclusively by the past ...', and two years later, when visiting the memorial at the site of the Bergen-Belsen concentration camp: 'We will always mourn in our hearts, not for everlasting enmity or fruitless hatred, but to gain strength and steadfastness.'[27] This echoes some German criticism of Israeli attitudes in the 1970s and seems to indicate that prominent Israelis now recognise that it is futile to concentrate their relationship with the Germans only on the past which, if it has left a legacy of deep sadness yet also offers challenges for the future.

NOTES

1. *Yediot Chadashot*, 8 May 1974.
2. *Jerusalem Post*, 22 June 1979.
3. *Rheinische Post*, 25 Jan. 1984.
4. There is still financial support for educational, charitable and similar institutions, both public and private, and much investment.
5. *Deutschland Berichte*, July/Aug. 1978.
6. *European Community Bulletin* 1980, no. 6.
7. *New York Times* (Weekly Review), 15 June 1980.
8. *Jerusalem Post*, 15 June 1980.
9. *Jerusalem Post*, 16 June 1980.
10. *Jerusalem Post*, 16 June 1980.
11. *Die Welt*, 16 June 1980.
12. *Der Spiegel*, 28 July 1980.
13. *FAZ*, 24 June 1978.
14. *Ma'ariv* quoted in above.
15. *Der Spiegel*, 3 July 1978.
16. *Der Spiegel*, 7 June 1982.
17. *Jerusalem Post*, 3 and 4 May 1981.
18. *Jerusalem Post*, 5 May 1981.
19. *Jerusalem Post*, 29 June 1978.
20. *Stuttgarter Zeitung*, 30 Jan. 1984.
21. *Der Spiegel*, 10 May 1982. The lifting of the arms ban occurred during the Falklands War.
22. *Stuttgarter Zeitung*, 30 Jan. 1984.

23. *Deutschland Berichte*, Dec. 1985.
24. Reported in *FAZ*, 10 April 1987.
25. *FAZ*, 19 Nov. 1987.
26. *Deutschland Berichte*, Nov. 1985.
27. *Deutschland Berichte*, May 1989.

12

Moral Debt or National Interest?

The thread running through the whole story of German–Israeli relations is without doubt the extermination of millions of Jews by the Hitler regime during the Second World War. This has deeply affected the attitudes to each other of the two governments and of public opinion on both sides. It is still affecting them half a century after the end of the war.

The Nazi crimes had left a legacy of hatred amongst many nations, some of which had suffered heavy losses of life, but by instituting what Hitler called the 'Final Solution to the Jewish problem', meaning the total extermination of the 'Jewish race', the Nazis had selected the Jews for special treatment. About one third of their number, men, women and children, were killed before an end was put to the massacre by the victorious Allied armies. Jewish people, including those living in Israel, have taken the view that the Germans must take some responsibility for the survivors. This explains the claim by the Israelis that the Germans owe them a moral debt and their demand that they are entitled to special consideration.

It is not easy to define a 'moral debt' in international politics. History does not record many examples of it. The implication must be that some country or people has inflicted serious damage or ill treatment on another and that it will voluntarily accept either liability for materially compensating the injured party or responsibility for its well-being in some other form. It must necessarily do so 'voluntarily' since there is no obligation in law. If it is not to be material compensation, then it must be help in some other way, such as for example diplomatic support. It can, of course, be both. But as nation states usually act from self-interest a conflict is certain to occur at some stage between moral debt and the national interest. If the moral debt is to be discharged in a way that is meaningful to the injured party, then the national interest must obviously take second place or be set aside altogether. Repaying a moral debt therefore requires some sacrifice on the part of the debtor country.

The story has been told in this book how Adenauer accepted the principle of compensation to Israel shortly after taking office. It is difficult to see how paying a very large amount of compensation to a state for damage caused before that state had even existed – especially when the debtor country is just recovering from economic ruin – can be attributed to the national interest. There would have to be either some advantage arising from it to the debtor country or strong external pressure. Neither was the case. That West German industry may have benefited by the delivery of goods to Israel under the Luxembourg Agreement, as has been suggested, was useful enough but cannot explain the motive. The money had to come from the West German exchequer, that is, ultimately out of the nation's earnings.

Adenauer was not under any pressure from the western powers, then occupying his country, to make such large payments to Israel. Indeed, the Americans were not helpful at any time from the moment they received Israel's compensation claim which they told the Israeli government to send directly to Bonn. The reasons for the American reluctance are clear: the US government was far less concerned about the former enemy than about a possible threat from a new one, the Communist Bloc. There were signs even before the founding of the Federal Republic that Washington was planning to boost the West German economy and this was reinforced later as a result of developments in the international situation. By the summer of 1952, when the Luxembourg Agreement was being negotiated, the Federal Republic had been given a minor but not unimportant role: that of participating in the defence of western Europe against possible aggression by the USSR. For this the Federal Republic was to be allowed to rearm. The young republic, still recovering from the ravages of war was, in common with other west European states which were in a similar situation, receiving American economic aid. Washington had reasons to worry lest the Federal Republic's economic recovery should be disturbed by a substantial drain of resources to another country or the Americans might be put into a position whereby they had ultimately to finance the large new debts which the West Germans might incur.

The credit for the successful outcome of the negotiations with Israel must therefore be given to Adenauer and some of his advisers who, in the face of resistance by the western allies as well as opposition among the German public and even by some of his cabinet colleagues, battled on until the treaty was signed. Though it may not be totally irrelevant that many West Germans were opposed to the treaty, this does not diminish the credit due to Adenauer and the West German people as a whole, for the treaty was ratified by the elected representatives of the Federal Republic. The Adenauer government and the two houses of the Federal Parliament recognised that there was a moral debt. They thus set a pattern that became an article of faith in the Federal Republic for nearly two decades and established what was agreed by both parties to be a special relationship. The Israeli side interpreted this as meaning that the Federal Republic had taken a measure of responsibility for the well-being and later also for the security of the Jewish State as atonement for the Nazi Holocaust. It explains Israel's request for weapons when it had difficulties in finding these elsewhere and Ben-Gurion's argument that Israel's chances had been diminished because the Nazis' murder of millions of Jews had deprived it of potential immigrants and it therefore needed more economic aid. The arms agreement is of special relevance in this context. It was a bold, indeed risky undertaking by Adenauer which had serious internal and external consequences. It could not be justified as an improvement to the Federal Republic's position in the Middle East: on the contrary, it put its role in the Arab world at considerable risk. It was not dictated by national interest but by a desire to help Israel. The same applies to the economic aid agreement of 1960.

A moral debt does not, of course, demand total submission. There were conflicts between the moral imperative and reasons of state where the latter carried the day. The most notable examples were the unwillingness of the Federal Government to establish diplomatic relations with Israel and the ending of the secret arms deliveries. The question of diplomatic relations would not have arisen in 1952 when the Federal Government would almost certainly have been willing to agree but the Israelis would not hear of it. By 1956 West Germany was a sovereign state and had devised its own foreign policy. The difficulty over diplomatic relations highlights two ingredients of the West German political scene which affected German–Israeli relations: one was the East–West conflict, the other the relationship between Israel and its Arab neighbours. The former had led to the division of Germany, but since 1955 the main objective of West German foreign policy had been reunification. The way the government handled this was virtually to refuse to acknowledge the division by ignoring and ostracising the German Democratic Republic, and reinforcing this tactic by the Hallstein Doctrine, devised to oblige the rest of the world to do the same. Thus Bonn's part in the East–West situation and its close relationship with Israel became mutually incompatible and the Federal Republic risked becoming indirectly embroiled in the Arab–Israeli conflict. The national interest in that case took over because the Federal Government was afraid first and foremost of ruining its chances of reunification, but also of losing its trade with, and unique position in, the Arab world. The ending of the secret arms agreement with Israel fell victim to the same difficulty.

Whether West German fears of Arab reactions to the establishment of diplomatic relations with Israel were justified is, of course, with hindsight, doubtful, but that is not the point. It is how governments perceive a situation at the time of acting that matters, and it is precisely mistakes in perception that can lead to conflict and even to war. In this case it led to anger in Israel when the arms deliveries were stopped. But there are at least some extenuating circumstances: neither the delay in diplomatic relations nor the abandonment of the arms supplies caused serious harm either to the security or the economy of Israel. Concerning the former, Ben-Gurion seems to have recognised this: he is believed not to have raised the matter of diplomatic relations with Adenauer at their meeting. At that time he was more interested in weapons and loans. His successor, Eshkol, accepted the end of the arms deliveries when Bonn offered financial compensation and weapons became obtainable elsewhere.

There were two other matters about which Israel felt aggrieved. One was Bonn's apparent inability to stop the West German scientists from participating in the build-up of the Egyptian president's war machine, the other was the Statute of Limitations which would have stopped the prosecution of war criminals and which Israel, in common with most of the rest of the world, therefore wanted to see abolished. In both cases the West Germans pleaded internal constitutional difficulties and these undoubtedly existed. But in the case of the scientists there was an extra complication, precisely the sort of misjudgment that can throw a relationship out of balance. The Israeli secret services had an exaggerated picture of what was really happening in Egypt. The bombastic utterances of the Egyptian government with threats of how they would destory the 'Zionist entity', designed to incite the masses in the Arab world, may have contributed to this error. But any Israeli political concern was also mixed with emotion because it was Germans who were threatening to finish off Hitler's job by destroying the survivors of the Holocaust in Israel. The Federal Republic, which owed the Israelis a moral debt, was vulnerable to accusations and reproaches if it did not take measures to stop its nationals creating weapons which threatened the survival of Israel. It was Prime Minister Ben-Gurion who played down the danger of the German contribution to the Egyptian weaponry to a realistic level. He had to balance the advantages of benefits such as the arms and economic aid which Israel was receiving against the only slight danger posed by the activities of the German scientists.

There remains at this time very considerable suspicion that the Federal Government procrastinated over acting against the scientists out of fear that Nasser might react by either recognising the German Democratic Republic or throwing the West Germans and their assets out of Egypt, as he threatened to do later. That would bring in the national interest as a motive for Bonn's behaviour. But at least a strenuous diplomatic effort is known to have been made behind the scenes. As for the Statute of

Limitations, that was an entirely internal West German problem with no disadvantage to anyone except that the handling of it brought severe criticism on the West Germans themselves. Precisely because it was a moral question whether the Germans were prepared to abandon the prosecution of war criminals and allow many of them to remain free, it was of great concern to Israel. It became a test case of whether the Germans were willing to overcome their Nazi past by abolishing the 20-year rule and in a sense a test of the sincerity of the moral debt to the Jews and Israel. In view of the prolonged opposition in the Federal Republic this test was not finally passed until the road was cleared for the continued prosecution of war criminals in 1979.

Changes in the perception of Israel by the West German people began in the late 1960s, shortly after the public elation during the Six Day War. A new attitude by the government became noticeable in the early 1970s when members remarked increasingly that there was no longer a 'special relationship' or that Israel was 'a state like any other'. It seems difficult to accept that there is still a special relationship between two states when one of them proclaims emphatically and often that it does not exist. The moral debt, which is at the root of the special relationship, certainly changed its quality at that time. The period when the Federal Republic was prepared to act outside the national interest, to make sacrifices or take risks because its government and people felt a moral imperative to help Israel, was over and was being replaced by something closer to political realism which argued that 'we will not forget the past but neither will we forego our freedom of action to suit Israeli policies or interests'. In practice they had not, of course, always done so. But in that event they were sufficiently sensitive to Israeli protests ultimately to come forward with a compromise that satisfied Israel. By their new approach the West Germans were signalling that the relationship was now normal and no longer special. Yet it cannot quite be said that Israel is to the Germans 'a state like any other'. When the existence of Israel was in danger during the Yom Kippur War the Federal Government, unlike its western European allies, came to its aid by quietly allowing the American airlift of arms to continue from its territory. More recently it has shown reticence, again unlike its neighbours, in sending arms to Israel's Arab enemies. These were not exactly sacrifices. The Brandt government took only a slight risk that its connivance at the American air lift might lead to an oil boycott. By 1987 the oil crisis, which had been the main incentive for European arms sales to Arab oil exporting countries, was over. But the fact is that the Federal Republic stood out among its European neighbours and decided to forego at least some advantages. That is an indication that there remains an awareness of a moral debt.

In Israel, it is true, the change in the West German attitude was not accepted. The Israelis had always insisted that material compensation could not make good the loss of millions of lives and that the moral debt was special and not subject to a time limit. This led at first to frequent protests by the Israeli government and media against, for example, West German government statements about the Palestinians and manifestations of friendship towards the Arab countries. During the 1980s there was concern in Jerusalem that the Federal Government, despite assurances that Israel must be allowed to live in peace within secure and recognised borders, was joining in what was called the 'Western arms bazaar' for the benefit of some Arab countries hostile to the Jewish State. The frequent Israeli admonitions were accompanied by reminders of past German treatment of the Jews and of the moral debt. These in turn caused irritation in the Federal Republic where the population increasingly grew tired of being reminded of the Germans' past. But there were signs also that during the 1980s Israel – Mr Begin's emotionalism apart – was coming to terms with the German change of attitude and that, although the past must not be forgotten, the future has to be faced. Israel has had to acknowledge that, whatever the criticisms voiced in the Federal Republic of Israeli policies in the Middle East, it is still obtaining a better deal from the West Germans than from almost any other European country. Among the important states of the world the Federal Republic, together with the USA, is still Israel's best friend.

The West German attitude to Israel has changed because memories of the past have faded with time, but also, and not least, because of changes in the Federal Republic's status in world politics. This had been largely the result of its rise as an economic power and of *Ostpolitik*, West Germany's partial reconciliation with the Soviet Bloc, which can now be seen as one stage towards the changes that have since taken place in eastern Europe. The division of Europe, a result of the Second World War, has come to an end and with it the division of Germany. But the end of the post-war era could lead to a further upvaluing of the status of Germany and a greater desire to draw a line under the past. It is undoubtedly the past which is holding the German-Israeli relationship together. There are still many people in the Federal Republic today who want to remember the past and are sympathetic towards Israel, and the government, even if it does not admit it, is still aware of the moral debt. But with the changes in Europe and the ultimate passing of the 'Holocaust generation' in both countries the relationship is likely one day to become normal in every sense.

Postscript

At the time of writing it is difficult to forecast what will be the future of the German–Israeli relationship. The old order, an international political system dominated by two rivalling superpowers, has come to an end. Germany – still known as the Federal Republic – has been reunified and Europe seems set to move towards closer integration in a community which may ultimately embrace most of the European continent, including some or all of the former communist states. Because of its size, its large population, its economic strength and, not least, its central position in a no longer divided Europe the new Germany now looks like becoming a dominant feature in the European Union. There are as yet few indications of what the German role will ultimately be or how the country will develop internally.

It has been suggested in previous chapters that the West German attitude to Israel changed, firstly because memories of the past were fading and a new generation was growing up, but also because of the Germans' increased self-assurance due to the Federal Republic's enhanced international status, which resulted from its economic strength and from 'Ostpolitik'. The new situation in Germany, created by unification, and its role as a potential leader in Europe could, but in the author's view need not, bring about a greater desire to draw a line under the past. If it does happen, then that might put an end to the moral debt.

There have been signs lately that the German government – still headed by Helmut Kohl – is abandoning the reticent foreign policy of the post-war years for a bolder approach. It has recently taken some initiatives such as, for example, the early recognition of the new Commonwealth of Independent States, which comprises some of the republics of the former Soviet Union, and also of two of the provinces of Yugoslavia which, in the course of a bitter civil war, have declared their independence. These steps were taken with only the reluctant approval of the other member states of the EU, which, however, followed suit. This led to criticism and in some European circles to the fear that the new Germany might weaken its links with the West and turn its attention more to Eastern Europe. The *Frankfurter Allgemeine Zeitung*, in a leading article, noted that Bonn has been criticised for being presumptuous and suspected of not being a reliable partner (in the EU) or ally (in NATO) and of trying ruthlessly to pursue is own interests. 'Bonn must,' the article continues, 'take account of this because it is understandable in the light of the history of this century and is nothing new ...' The *New York Herald Tribune*, published in Paris, referred to the dismay of western governments at the new self-assertiveness of German diplomacy which, it believes, is bound to increase concern about the risk of an independent, nationalistic German policy. Fears that Germany might go it alone were previously expressed during the early 1970s in the days of *Ostpolitik*. These and other criticisms have served as a reminder that, whatever may be happening in Germany, the rest of the world has not forgotten the past.

So far there is no sign of a change in German policy towards Israel. There was a test of this during the recent Gulf War resulting from the attack by Iraq on Kuwait. It was reported that German firms had delivered chemical and biological weapons to Iraq, prior to the invasion of Kuwait which triggered the United Nations' action. This caused some embarrassment to the Kohl government. Israel, which was not involved in the campaign against Iraq and was prevailed upon by Washington not to intervene, was nevertheless attacked by Iraqi Scud missiles, procured over the years from the Soviet Union. It was feared in Israel and other countries that these might contain chemical or biological warheads. They did not, but the Federal Government made the gesture of offering financial compensation to Israel, some of which has been used for the purchase of armaments. The export of chemical, biological and other weapons is illegal in Germany and the government has since instituted legal proceedings against the offending firms.

Internal political developments in the new Germany have been less reassuring. The collapse of the USSR and the dissolution of the Soviet Bloc, coupled with changes in all the former member states from a communist political and economic system to democracy and a market economy, has created upheaval, disorientation and insecurity in all the affected states. Germany is, of course, deeply involved because what was the German Democratic Republic effectively became part of West Germany. The Bonn government is largely financing the change-over in East Germany, where factories and businesses, which are unable to compete in a market economy, have to be reorganised or in some cases permanently closed. But the formal reunification in October 1990 was timed to precede a Bundestag election by approximately two months. It is most unlikely that the government of Helmut Kohl would have won this election without the euphoria of reunification and without unrealistic government promises of a quick upturn of the economy in the areas of the former GDR. By early 1993 Germany found itself in the depth of a world-wide economic recession, with the eastern Länder bearing the brunt of the suffering in the form of mass unemployment, a general decline in living standards and a bleak outlook for the future, which leaves many East Germans helpless and frightened.

But there is also discontent in the western parts, which had been used to unprecedented prosperity. Here, too, economic activity has sharply declined and unemployment has risen to unaccustomed levels. In both parts of the country discontent has been increased by the large number of foreign immigrants, some of whom have lived in western Germany for 30 years or more, having been attracted by the shortage of labour during the economic boom in the 1950s and 1960s. Others have come in recently as refugees from various trouble-spots, including the formerly communist countries of eastern Europe, the Middle East or even from as far as southern and south-eastern Asia. The liberal German immigration laws, embodied in the Basic Law after the Second World War and changed only very recently because of public pressure, had allowed this large influx and have – not unnaturally – affected the employment market, the availability of housing and public finance.

It is recalled here that during a brief recession in the 1960s the economic uncertainty led to an increase of support for extreme right-wing and neo-Nazi movements. At that time this phenomenon disappeared rapidly once the crisis was over. There is not much evidence at present that right-wing extremists are losing ground. But the recession this time is deep and lasting a great deal longer. In some circles it has led to manifestations of 'weariness of politics' (*Politikmüdigkeit*) and a distrust of politicans and of the political system, sometimes leading to violence. There have, for example, been vicious attacks by neo-Nazis in both parts of Germany on hostels for foreign immigrants, in which residents have been killed or injured. There have also been vociferous demonstrations by small groups of neo-Nazis extolling the Hitler régime and shouting anti-Semitic slogans or desecrating Jewish cemeteries, despite the fact that there are fewer than 50,000 Jews living there now in a population of 80 million. It seems incredible that after West Germany had experienced over 40 years of successful parliamentary democracy together with growing prosperity, some of its citizens should want to return to a period of chauvinistic dictatorship with all its attendant loss of life, destruction and misery. Although some right-wing parties have recently entered Länder parliaments, their voting strength has fortunately been slight so far and nothing like sufficient to constitute a danger to the parliamentary system. But the warning signs are there.

Fortunately, these people only constitute a very small minority of the population. Early last year the fiftieth anniversary not only of the end of the Second World War but also of the destruction of Auschwitz extermination camp were observed by many people and much could be read about the Holocaust in the newspapers. It showed that a very large proportion of the German population did not want to forget.

It would be disastrous for German–Israeli relations if anti-Semitism, which tends to flourish in such situations, came to the fore again in Germany. In that case, that is if a neo-Nazi or other similarly anti-Semitic group were to grow into even a substantial minority party, it would confirm to a large part of the Israeli population what it had tried to forget during the past 40 years – that the Germans had never changed and could not be trusted.

For Israel the new world order has created some problems but also opportunities. The demise of the Soviet Union and the end of superpower rivalry in the Middle East has meant that Soviet support for Israel's Arab opponents, especially in the form of the supply of modern weapons, has ended, while the United States of America remain friends of Israel. This has encouraged some of Israel's Arab neighbours, which were still technically at war with the Jewish State, to negotiate about an end to hostilities. In the case of the Hashemite Kingdom of Jordan these negotiations have been successful and at present rather more difficult talks are taking place with Syria. Among the more distant Arab states Morocco and Tunisia and some of the Gulf States have established diplomatic relations or are about to do so. All this has been made easier by the fact that a left-of-centre coalition government in Israel (led at first by Prime Minister Yitzhak Rabin, who sadly was assassinated by an extremist of his own people), taking advantage of the changed circumstances, has agreed to negotiate with the leader of the Palestine Liberation Organisation, Yassir Arafat, about giving a measure of autonomy to the Palestinians in the Israeli occupied areas, starting with the Gaza Strip and the city of Jericho and now extended to other areas.

This initiative, which is actively supported by the United States and Egypt, Israel's first Arab neighbour to have made peace more than a decade ago, will hopefully lead eventually to Israel being accepted by all Arab states and thus to peace in the Middle East. The main stumbling block at this moment is religious fundamentalism, on both the Islamic and Jewish sides, using terrorist methods as a means of frustrating the efforts for peace. If this can be overcome, that would also mean the end of Israel's economic isolation and obviate for Israel the necessity of spending an excessively large share of its GNP on weapons. It would reduce its dependence on foreign aid, especially that coming from its main benefactors, the United States and the Federal Republic of Germany.

G.L.
February 1996

Bibliography

Abediseid, M., *Die deutsch-arabischen Beziehungen*, Stuttgart, 1976.

Adenauer, K., *Erinnerungen*, vol. 1, 1945–53, Frankfurt, 1967.

—, *Erinnerungen*, vol. 2, 1955–59, Frankfurt, 1969.

—, 'Bilanz einer Reise – Deutschlands Verhältnis zu Israel', in *Die politische Meinung*, vol. 11, no. 115 (1966).

Balabkins, N., *West German Reparations to Israel*, New Brunswick, 1971.

Ben-Natan, A., *Briefe an den Botschafter*, Frankfurt, 1971.

Ben-Vered, A., 'Deutschland und Israel. Bedeutung der Aufnahme von diplomatischen Beziehungen für den jüdischen Staat', in *Europa-Archiv*, 1965, no. 13.

—, 'Über die Entwicklung der deutsch-israelischen Beziehungen', in *Europa-Archiv*, 1967, no. 7.

—, 'Die Rolle Israels in der Weltpolitik' in *Mittlere Mächte in der Weltpolitik* (Schriftenreihe des Forschungsinstituts der Deutschen Gesellschaft für Auswärtige Politik), Opladen, 1969.

Birrenbach, K., 'Die Aufnahme der diplomatischen Beziehungen zwischen der Bundesrepublik Deutschland und Israel', in *Ludwig Erhard, Beiträge zu seiner politischen Biographie*, Frankfurt: Festschrift, 1972.

Böhm, F., 'Die deutsch-israelischen Beziehungen', in *Frankfurter Hefte*, vol. 20, no. 9 (Sept. 1965).

—, 'Ein Staat wie jeder andere?', in *Tribüne*, vol. 9, no. 33.

Büttner, F., 'German Perceptions of the Middle East Conflict: Images and Identifications during the 1967 War', in *Journal of Palestine Studies*, vol. 6, no. 2 (1977).

Cygielman, V., 'Trusting Bonn again', in *New Outlook*, vol. 8, no. 2 (1965).

Dawidowicz, L., *The War against the Jews 1933–1945*, London, 1975.

Deligdisch, J., *Die Einstellung der Bundesrepublik Deutschland zum Staate Israel*, Bonn-Bad Godesberg, 1974.

Deutschkron, I., *Israel und die Deutschen. Zwischen Ressentiment und Ratio*, Cologne, 1970.

Dittmar, P., 'Die DDR und Israel. Ambivalenz einer Nichtbeziehung', in *Deutschland Archiv*, vol. 10, no. 2 (1977) (2 articles).

Eban, A., 'Bestandteile der Beziehungen zwischen der Bundesrepublik Deutschland und dem Staat Israel', in *Diplomatischer Kurier*, vol. 19, no. 6 (1970).

Evron, Y., *The Middle East. Nations, Superpowers and Wars*, London, 1973.

Feldman, L.G., *The Special Relationship between West Germany and Israel*, London, 1984.

Frei, N., 'Die deutsche Wiedergutmachungspolitik gegenüber Israel im Urteil der öffentlichen Meinung der USA', in L. Herbst & C. Goschler (eds.): *Wiedergutmachung in der Bundesrepublik Deutschland*, Munich, 1989.

Frei, O., 'Aussenpolitische Anstrengungen der DDR in der nicht-kommunistischen Welt', in *Europa-Archiv*, 1965, no. 22.

Gerlach, F., *The Tragic Triangle. Israel, Divided Germany and the Arabs, 1956–1965*, PhD Thesis, Columbia University, 1968.

Gegenwartskunde: 'Die BRD zwischen Israel und den Arabischen Staaten', (leading article), no. 1, 1975.

Goldman, N., *The Autobiography of Nahum Goldmann. Sixty Years of Jewish Life*, New York, 1969.

Gollwitzer, H., 'Zur Frage der deutsch-israelischen Beziehungen: Ein Ueberblick: Zu dem offenen Brief deutscher Hochschullehrer', in *Blätter für deutsche und internationale Politik*, 1964, vol. 9, no. 12.

Grosser, A., *Germany in Our Time. A Political History of Post War Years*, London, 1974.

Haftendorn, H., *Militärhilfe und Rüstungsexporte der Bundesrepublik Deutschland*, Düsseldorf, 1971.

Hottinger, A., 'Die Hintergründe der Einladung Ulbrichts nach Kairo', in *Europa-Archiv*, 1965, no. 4.

—, 'Die arabische Welt zwischen der Israel-Front und der "Erdölwaffe"', in *Europa-Archiv*, 1973, no. 7.

—, 'Der vierte arabisch-israelische Krieg und seine Folgen', in *Europa-Archiv*, 1974, no. 3.

Imhoff, C., 'Neue Orient-Politik des Generals de Gaulle', in *Aussenpolitik*, vol. 19, no. 1 (Jan. 1968).

—, 'Die konstruierte Krise', in *Tribüne*, vol. 9, no. 33 (1970).

Jena, K. von, 'Versöhnung mit Israel? Die deutsch-israelischen Verhandlungen bis zum Wiedergutmachungsabkommen von 1952', in *Vierteljahrsheft für Zeitgeschichte*, vol. 36, no. 4 (Oct. 1986).

Kissinger, H., *Years of Upheaval*, London, 1982.

Kreysler, J. und Jungfer, K., *Deutsche Israel-Politik. Entwicklung oder politische Masche?*, Diessen, 1965.

Laqueur, W., *The Soviet Union and the Middle East 1958–68*, London, 1969.

Lewan, K., *Der Nahostkrieg in der westdeutschen Presse*, Cologne, 1970.

Medzini, M., 'Israel's Changing Image in the German Mass Media, in *Wiener Library Bulletin*, vol. 26, no. 3/4 (new series nos. 28/9, 1972/73).

Pauls, R., 'Aussenpolitik und Entwicklungshilfe', in *Aussenpolitik*, vol. 16, no. 6 (June 1965).

Pulzer, P., *The Rise of Political Antisemitism in Germany and Austria*, London, 1988.

Rolef, S.H., *The Middle East Policy of the Federal Republic of Germany*, Jerusalem, 1985.

Seelbach, J., *Die Aufnahme der diplomatischen Beziehungen zu Israel als Problem der deutschen Politik seit 1955*, Meisenheim, 1970.

Serre, F. de la, 'L'Europe des Neuf et le conflit israélo-arabe', in *Revue française de Science politique*, vol. 26, no. 4 (1974).

Shinnar, F., *Bericht eines Beauftragten. Die deutsch-israelischen Beziehungen 1951–66*, Tübingen, 1967.

Sontheimer, K., *Die verunsicherte Republik. Die Bundesrepublik nach 30 Jahren*, Munich, 1979.

Steinbach, U., 'German Policy on the Middle East and the Gulf', in *Aussenpolitik* (English edition), vol. 32, no. 4 (1981).

Vogel, R., *The German Path to Israel*, London, 1969.

— (ed.), *Der deutsch-israelische Dialog. Dokumentation eines erregenden Kapitels deutscher Aussenpolitik*, Teil 1: 'Politik', vols. 1–3, Munich, 1987/88.

Wagner, W., 'Untersuchung eines deutschen politischen Instrumentariums', in *Europa-Archiv*, 1965, no. 21.

—, 'Rückschläge der Bonner Politik in den arabischen Staaten', in *Europa-Archiv*, 1965, no. 10.

Well, G. van, 'Die Entwicklung einer gemeinsamen Nahost-Politik der Neun', in *Europa-Archiv*, 1976, no. 4.

Wewer, H., 'Israel und das politische Selbstverständnis der Deutschen', in *DISkussion*, vol. 7, no. 18 (1966).

Wolffsohn, M., *Ewige Schuld? 40 Jahre deutsch-jüdisch-israelische Beziehungen*, Munich, 1988.

—, 'Das deutsch-israelische Wiedergutmachungsabkommen von 1952 im internationalen Zusammenhang', in *Vierteljahrsheft für Zeitgeschichte*, vol. 36, no. 4 (Oct. 1988).

—, 'Globalentschädigung für Israel und die Juden? Adenauer und die Opposition in der Bundesregierung', in L. Herbst & C. Goschler (eds.), *Wiedergutmachung in der Bundesrepublik Deutschland*, Munich, 1989.

Index

German scientists, 59, 61–5, 70, 82, 86,
90–1, 93, 101, 105, 112, 115, 117–18,
120, 125, 141, 152, 208
Israel (conflict with), 30, 42, 66, 99, 124,
193, 197, 208, 214
non-alignment, 59, 109
Suez crisis, 42
West Germany, 38, 43, 103–5, 107, 208;
economic aid, 43, 94, 100, 108–9, 114,
123, 125
Yemen campaign, 98–9, 123, 148, 151
Yom Kippur War, 178–9
Eichmann, Adolf (trial), 45, 61, 65, 77, 87,
157
Eilat, 147, 155
Eisenhower, Dwight, 54
Erhard, Ludwig, 11, 56–7, 78–9, 82, 86,
93–6, 103, 105–8, 111–18, 120–2, 126,
130, 132, 134–6, 142
Erler, Fritz, 86
Eshkol, Levi, 93–4, 104–6, 108, 113, 115,
117, 130, 134–5, 138–9, 147, 208
European Communities (EC), European
Union (EU), 164, 184–5, 197
Coal and Steel (ECSC), 5, 25
Defence (EDC), 7–9
Economic (EEC), 38, 124, 202
joint foreign policy approach, 170–4
Middle East declarations, 184–5, 195–6
Middle East report, 170–1, 185
European Union, 211
'extra-parliamentary opposition' (APO)
see New Left

Fahmy, Ismail, 179
Federer, Georg, 100
Final Solution (of Jewish question), 9, 76–7
France
Arab dependencies, 30, 124
arms for Israel, 31, 48, 182
de Gaulle, 124, 144, 148, 159, 164
Germany (West), 167
relations with Arabs, 173
Suez crisis, 37
Free Democratic Party (FDP), 45, 80, 84,
120, 126, 151, 165, 168

Gaza Strip, 147, 194, 214
Genscher, Hans-Dietrich, 199
German–Israel Society, 153, 181
Germany, pre-1949, 2, 4, 7
division, 106, 207
Hitler period (Third Reich), 3–4, 6, 9, 17,
25, 72, 74, 76, 79, 88, 91, 131–2, 151–2,
159, 160, 205
occupation, 1, 4, 21

Potsdam Agreement, 1, 3, 81, 83, 137
refugees, 8
war victims, 8
Germany, Federal Republic, 105, 125–6,
164, 166
Arab countries, 12, 24, 31, 96, 101, 103,
105, 120–2, 124, 175, 207; arms
exports, 43, 51, 200–2, 210; diplomatic
relations, 94, 149, 169; economic aid,
124; oil boycott, 178
arms exports to Middle East, 201
Basic Law (constitution), 67, 70, 82, 85
church organisations, 153
Claims Conference, 11; contractual
agreements, 7
compensation to Israel for arms stop, 106,
112
economic recession (1966), 142–3
Egypt, 105–6, 108, 122
establishment and legitimation, 3–4
European integration, 5, 8, 11, 21, 25,
167
former Nazis, 27, 45, 56, 59, 74–5
Grand Coalition, 142–3, 146
Hallstein Doctrine, 22, 28, 33, 35–6,
43–6, 90, 96, 97–8, 105, 107, 109–10,
114, 121, 124–6, 169–70, 206
illegal export of biological and chemical
weapons from Fed. Rep., 212
immigration, 213
Israel, arms deliveries, 49, 56–7, 88,
90–2, 94–5, 101–8, 115–17, 120–1,
125, 134; arms purchases, 49;
diplomatic relations, 27, 31, 33–5, 39,
46, 90–1, 94–5, 101–4, 112–15, 117–21,
124, 141, 207; economic aid, 55, 90,
101, 105, 107, 116–17, 122, 125, 135–6,
161, 207; public attitudes in Israel to
Germans, 13, 63, 106, 108, 118–19,
159; public attitudes in Germany to
Israelis, 89, 159–61, 187–8, 200; trade,
202; visits, 175
Jews, 1, 5, 12, 120, 155, 174, 205
NATO, 22, 25–6, 37, 56, 167
occupation, 4, 21
overcoming of past, 52, 54, 72
reunification, 29, 35, 126, 135–6, 165–7,
207, 210–11
recession in reunited Germany, 213
sovereignty, 22–3, 29, 207
Students' Federation, 156
Trade Union Federation (DGB), 45, 120
Germany, German Democratic Republic,
6, 76–8, 127, 212
Berlin, 97
diplomatic recognition, 22–4, 34, 39,